ENVIRONMENTAL ACTION

Environmental Action is written by members of the environmental team at Leigh, Day & Co. Solicitors and is edited by the head of the team Martyn Day. The members of the team contributing to the writing of the book are:

> Gisele Bakkenist
> Alan Care
> Martyn Day
> Claire Hodgson
> Charles Hopkins (to December 1996)
> Sean Humber
> Sheena Lee
> Richard Meeran
> Sally Moore
> Suzanna Read
> James Schuman
> Richard Stein

ENVIRONMENTAL ACTION

A Citizen's Guide

Edited by Martyn Day

Pluto  Press
LONDON • CHICAGO, ILLINOIS

First published 1998 by Pluto Press
345 Archway Road, London N6 5AA
and 1436 West Randolph, Chicago, Illinois 60607, USA

British Library Cataloguing in Publication Data
A catalogue record for this book is available from
the British Library

ISBN 0 7453 1191 1 (hbk)

Library of Congress Cataloging in Publication Data
Environmental action : a citizen's guide / edited by Martyn Day.
    p.   cm.
  ISBN 0–7453–1191–1 (hardbound)
    1. Environmental law—Great Britain. 2. Liability for
environmental damages—Great Britain. 3. Citizen suits (Civil
procedure)—Great Britain. I. Day, Martyn.
KD3372.E579   1997
344.41'046—dc21   97–22286
                                          CIP

Designed, typeset and produced for Pluto Press by
Chase Production Services, Chadlington, OX7 3LN
Printed in the EC by TJ International, Padstow

Contents

Preface

In March 1984 the Labour MP Frank Cook asked if I would join a group he was just setting up called the Radiation Victims' Roundtable. The group, which consisted of scientists, environmentalists, politicians and lawyers, was devoted to doing what it could to help those whose lives had been impacted by radiation. My work on the Roundtable made me realise how poorly supported are people fighting to keep their environments secure in this country.

As a result of this work I was asked by a family, called the Merlins, to work on their case, which was a claim for the financial losses resulting from the impact on their property of their living on an estuary just south of the Sellafield nuclear reprocessing plant. They had been unable to sell their property when it became apparent that radioactive waste from the sea, emanating from the plant, was coming back into their home. We fought that case and were swamped by the defendants and their lawyers, who had devoted millions to fighting the case, in contrast to the minute fraction that *we* had spent in the pursuit of the claim.

When the judgment against the family was given in April 1990, I committed myself and my firm to ensuring that when it came to environmental cases anyone instructing Leigh, Day & Co. would receive the same level of service as their opponents would from the huge City law firms. As a result, I have brought together a team of lawyers who are devoted to protecting the environmental rights of the ordinary citizen.

This book is a part of that process. It has been written by the lawyers who have helped form the Environmental Department at Leigh, Day & Co. and it draws deeply on our accumulated experiences of the last thirteen years.

The lawyers who have contributed to the writing of this book are:

Gisele Bakkenist
Alan Care
Martyn Day
Claire Hodgson
Charles Hopkins
Sean Humber
Sheena Lee
Richard Meeran
Sally Moore
Suzanna Read
James Schuman
Richard Stein

The secretaries in the department deserve a special word of praise for their devotion to the project, being Michelle Tudor, Bridget Heapy, Shirley Bright, Connie Durrant and Linda Marlow.

I am very grateful to Fiona Reynolds, Adam Woolf, Sandhya Drew, Richard Parkes and Andy Radford for their invaluable ideas on improving the contents and for Alan Watson and Julie Ratti for going through the book with a fine-tooth comb, putting us right in innumerable areas. However, any mistakes or omissions remain the exclusive property of the authors. At Pluto, Anne Beech and Robert Webb have given us great support for which I am grateful.

The book has been completed in the immediate aftermath of New Labour's election victory and with the hopes for a new dawn in many policy areas, including the environment. It would be a pleasure if, as a result of this change, this book were to soon become obsolete. Somehow I doubt it.

Martyn Day
July 1997

1 Introduction

WHY TURN TO THE LAW?

This is a book that is designed for and dedicated to all those citizens within our society who are interested in the environment and who are prepared to take steps to protect it. It is a book about action, about hope and above all about giving people the weapons with which to fight their corner in all sorts of environmental battles. It is a book about the empowerment of the citizen.

In the last ten years Britain has seen an explosion of environmental actions taken by people at all sorts of levels in all sorts of ways. The membership of established organisations like Greenpeace, Friends of the Earth and the Council for the Protection of Rural England has shot up as people have wanted to show their support for the environmental causes these groups espouse.

Similarly, new single-interest environmental groups have sprung up all over the place, with Reclaim the Streets and Surfers Against Sewage being prime examples.

We have entered an era when citizens are no longer prepared to stand idly by while their environment, their neighbourhood and their quality of life are damaged by the interests of big business, government or even their local councils. People are increasingly turning to existing environmental groups to join forces in trying to ensure that the environment is protected, not just for today but for the next generation and for many generations to come.

Individual citizens are not only showing their concerns by joining environmental groups. Increasingly they are proving their readiness to take their own action to protect their environment. This action can take the form of demonstrating, lobbying, letter writing, or appealing to the media, but it has also included taking legal action.

In the last decade citizens have increasingly turned to the law in an attempt to find ways to prevent environmental damage or to obtain redress for damage that has already occurred. The aim of this book is to explain, as comprehensively as possible, how the law operates in relation to all sorts of actions, whether they are taken by just one individual, an organised group of individuals or an environmental organisation, and whether they are taken against an individual, a company, or even a local authority.

Although the law is always unpredictable and can seem like an obstacle course (even to those who spend their working lives involved in pursuing legal actions), it can also be a great leveller. At times, it is the one way of giving the citizen the power to tackle, on an even playing field, those mighty corporate bodies who damage or propose to damage the environment. The threat of court action cannot be shrugged off by a chief executive whose company faces a realistic prospect of legal defeat.

It should be remembered that all the uncertainties and hurdles faced by citizens taking legal action are also problems for the opponents in a legal case.

For many people the concept of turning to the law is a daunting one, and the prospect of consulting a lawyer holds all the appeal of a wisdom tooth extraction. This book should give its readers the confidence to treat the law as an ally rather than an enemy. For those contemplating environmental action, it offers a formidable array of legal weaponry that can be wielded in the battle to be waged. The law may not always have an answer at least, not always an affordable answer but it is an aspect of environmental actions that must be kept firmly in mind.

THE BOOK

Many people reading this book will be doing so because they have become involved in an action of one sort or another, and will need to use it as a reference to help them understand the legal process. The authors have therefore tried to ensure that each chapter is complete in itself, as far as that is possible. It is inevitable, however, that one chapter will touch on areas that are explored further in another chapter, and when this happens cross-references are provided.

The book has also been written to help individuals who are involved in environmental issues to understand their rights and how the legal process works. Although this inevitably means that the book considers campaigning issues it is not a campaigner's handbook. Advice of that kind is best obtained from national environmental groups, such as Greenpeace, Friends of the Earth and the Council for the Protection of Rural England.

In producing a legal book for the individual, the authors have had to strike a balance, providing adequate information without being overly technical. Their intention is to offer enough legal detail to give a clear understanding, without burdening the reader with unnecessary material, such as specific references. For example, where cases already determined by the courts are discussed, the name of the case is given without the full court reference. Also, because this is a book on legal rights, legalese is occasionally unavoidable. To assist the lay person, a glossary of legal terms is set out at the end of the book.

This book is an edited collection. The authors, while being from the same law firm, are individuals with distinctive approaches, and all are experts in the fields they have written about. Because of the varied nature of the subjects, some chapters are unavoidably more technical than others. In certain areas, someone facing the prospect of legal action will almost certainly instruct lawyers in the case; those chapters are more discursive and can be used to assist in following the action. There are other topics, such as planning, where readers may well be attempting to take action without obtaining legal representation; those chapters give much more technical detail.

THE STRUCTURE

The book is divided into three parts. The first part deals with the primary issues involved in preparing for action. The second looks at what a community can do to stop a development that it opposes. The third part considers how to make various types of claim where there is a development that already exists.

Part 1: Preparing for action

First steps

Most environmental actions only commence once local people
have become concerned about an issue and have joined forces.
Occasionally individuals are prepared to take this step alone, but
in most cases people look for strength in numbers by forming a
new group or by bringing in an existing organisation to help.
Chapter 2 considers how people join together in groups. It takes as
its example a group of people in Wales who formed to take action
against the operation of one of the plants of the chemical giant,
Monsanto. The chapter offers ideas and warnings for those
contemplating such a step.

Funding

One of the key issues facing anyone who becomes entangled with
the law is cost. Where there are criminal charges against someone
involved in a direct action there is the possibility of obtaining
legal aid where the person's means are low enough.

Usually, there are lawyers who are prepared to represent people
involved in such actions, whether for free or under the legal aid
scheme or at some agreed rate. The environmental legal group
Earthrights, for example, organised the representation for campaigners arrested and/or charged in the Newbury Bypass direct
action. Costs are not normally a major issue in these cases, not
least because the charges are usually relatively minor and the
resources put into them by the police and Criminal Prosecution
Service are by no means huge.

However, costs are much more of an issue in civil proceedings.
In this type of legal action in Britain, costs primarily follow the
event, which means that the losing party pays not only his/her
own costs but also the costs of the winners. That is of great value
to the winners, but it means that anyone commencing a civil
action has to be clear that losing the case may prove to be
extremely expensive.

This rule does not, in practice, apply in cases where the plaintiff or
defendant is legally aided (the 'plaintiff' being the person or organi-

sation who brings a civil action that is, the maker of the claim and the 'defendant' being the person or organisation against whom the claim is made). The problem here is that the parameters for people being able to gain legal aid are becoming increasingly narrow. Chapter 3 explores this issue, explaining how the legal aid system works, and under what circumstances a claimant will be able to obtain a legal aid certificate, which gives some comfort of knowing that the plaintiff's costs will, in most cases, be paid by the Legal Aid Fund, in whole or in part, rather than out of the claimant's pocket.

In most circumstances the defendant's costs, if a plaintiff with legal aid loses a case, will be borne by the defendant. An order may be made against the plaintiff but this will only usually be enforceable with the court's specific approval, which is normally given only where that has been a dramatic improvement in the plaintiff's financial position. Such an order is often referred to as being a 'football pools order', for obvious reasons (these days it might be more appropriate to call it a 'National Lottery order').

It is the people who are on middle incomes who have greatest difficulty in funding an action through our legal system. The rich can afford to pay for legal representation and the poor can obtain legal aid. But legal bills that can easily run to £50,000–£100,000 are beyond the means of those in the middle-income bracket.

Going to the law is never cheap, with the costs of the barristers, experts, the court itself, and the solicitors usually adding up to a substantial total. However, environmental actions can be particularly costly. The defendants in these actions are often some of the largest corporations in the country, if not in the world, and so the legal fight is likely to be of great importance to them, not least in terms of public relations. The companies, therefore, will be prepared to spend vast sums in defending actions. BAT, the tobacco giant, announced at its 1996 Annual General Meeting that it had spent, in the previous year, some $60 million defending legal actions by smoking victims, primarily based in the US, and yet this was only a fraction of its overall legal bill for commercial work.

The media
The media are one of the campaigner's greatest assets and Chapter 4 offers some practical suggestions about how they might be used

to gain support for the aims of the group. One of the best examples of successful use of the media is perhaps that of the *Brent Spar*, where the combined efforts of Greenpeace, the media and public opinion, particularly in Germany, brought about Shell's stunning climb-down in agreeing not to dump the oil rig in the sea.

Environmental information

One of the primary concerns for most local campaigners is the lack of information about environmental developments that are taking place. There is increasing European legislation about this and Chapter 5 explains what information must be held, where it is held, and how accessible it is. This is an area where there have been a number of legal actions that have tried to open up an instinctively protective corporate world with a modicum of success. It is an area where the concerned citizen should certainly contemplate legal action.

Europe

With so many years of government enthusiasm toward corporate developments of any sort, and with few checks on the onward march of big business, the environmental movement opposing these changes has increasingly looked to Europe for the odd crumb of comfort. In the final part in this section, Chapter 6 considers the issue of Europe, explaining how it operates, what sort of case might end up there and, generally, how a development might be caught up in the European Union's systems and structures. There is always the tendency to look at the mechanisms of the EU and the European Court of Justice as if they were from another planet, or alternatively to consider that any and every case is ripe for a European challenge. Neither position is correct and an understanding of the process in Brussels, Strasburg and Luxembourg can be very beneficial to the campaign.

Part II: Actions to stop new developments

When it comes to making an effort to stop environmental damage occurring there are various different actions that can be taken, including use of the law. When a new development is proposed, those opposing have a number of alternatives when trying to stop the action going ahead. The first is intervening in the planning process; if that is unsuccessful, there is the possibility of judicially reviewing decisions to allow the development to go ahead; finally, there are public demonstrations against the development. These routes are by no means mutually exclusive and they can all be brought to play in attempts to prevent a development taking place.

The first two routes use the civil law, which primarily regulates relationships and competing rights between groups in society. The third route is likely to bring the demonstrators up against the criminal law, which is designed to enforce those rules considered to be the most important in terms of governing the conduct of individuals and organisations within our country, with the state prescribing the punishment to be levied for a breach of those rules.

Planning and road proposals

The primary way of trying to prevent environmental damage occurring is through intervention at the planning stage. Although this is nothing like as exciting as a good demonstration or other form of direct action, it can have real and lasting results provided the necessary work is put in to achieving the aims of the campaigners.

The planning system in this country is extremely sophisticated, with a whole structure surrounding it and the campaigner intent upon taking the fight to the developer during the planning process must take the time to understand how the process operates and to identify the most opportune times for an intervention.

Chapters 7 and 8 on planning and new road proposals look at this issue in some depth, not least because this is an area where objectors may well be able to put forward the objections themselves rather than bring in lawyers with all the incumbent concerns about costs.

The purpose of these chapters is to give objectors sufficient information to allow them to represent themselves at the planning enquiry rather than use experts. The chapters therefore offer considerable detail about the system's structure and process. Of course, there may well be circumstances where it would be worth the objectors trying to find the funds to bring in professionals, but this may simply not be possible. Besides, in many cases the objectors could do as good a job.

Judicial review

Another route for trying to prevent environmental damage taking place is through the use of the courts by way of judicial review. This is a legal tool that has been used increasingly in recent years and there are now journals devoted to environmental judicial review cases, where the courts inquire whether an administrative decision has been taken in a fair manner given the circumstances pertaining at the time.

The principle underlying these proceedings is that they are requests to the courts to overturn decisions allowing developments to take place where these are thought likely to lead to environmental damage. This administrative law remedy has had various levels of success and failure but is clearly a tool that should be considered by anyone wanting to stop a development taking place.

Examples of judicial reviews in recent years are:

- a review of the government's decision to fund the building of the Pergau Dam in Malaysia;

- a review related to the installation by the National Grid of a new underground 400 kilovolt power line in north London;

- the review by Greenpeace of the decision giving authority for the Thorp Nuclear Reprocessing Plant at Sellafield to commence its operations;

- the review by local children of the refusal by Greenwich Council to close its main road on days when the air quality is at its worse.

Chapter 9 looks carefully at the circumstances when judicial reviews can be brought, who can bring them and the extent to which a victory in the courts is likely to be a knockout blow in favour of the citizen.

A prime example of where this was not the case was in relation to the judicial review by the pressure group Save Our Railways, who reviewed an aspect of the government's privatisation of the rail network. The Court of Appeal gave judgment largely in favour of the group in December 1995 but within a matter of weeks the government had simply changed the rules regarding the privatisation in a way that allowed them to make the same decision that had been reviewed. The decision undoubtedly caused the government major embarrassment, as the case was used to damage the public relations exercise surrounding the privatisation. It is significant, though, that because the government has the ultimate power of legislative change, it can, in effect, sidestep a judicial review defeat.

Despite that, there have been a number of occasions when judicial reviews have been able to provide a knockout blow against the decision-maker. Nevertheless, a citizen or a group planning to use judicial reviews should be clear about their limitations.

Public protest

The number of instances where people have been prepared to make a public protest to try to prevent environmental damage occurring has dramatically increased in recent years. The organisation that personifies this type of action at both national and international level is Greenpeace, whose victory over Shell in relation to the decommissioning of the Brent Spar platform in the North East Atlantic, and the pressure applied against the French government's nuclear testing in the South Pacific, showed how a well-motivated and well-structured organisation could have a dramatic impact on the world stage.

Examples of more local actions are those by Reclaim the Streets in London, who have blocked roads during the rush hour by occupying key junctions, and by people who have been willing to risk their lives in an attempt to stop the building of the M11 extension and the Newbury Bypass.

These direct actions can be organised simply by one group, such as those by Greenpeace, or there can be an amalgam of groups and individuals, such as was the case with the Newbury Bypass campaign.

In taking direct action the main legal concern is what is likely to happen to those individuals and groups in the courts as a result of the actions of the police or those against whom the protest is directed. Chapter 10 on public protest looks at the criminal charges that can be brought against individuals and groups, at the use of injunctions (court orders requiring named persons to refrain from carrying out certain acts) and at contempt charges that can be brought against those breaching injunctions in disobedience of the court.

The uncertainties underlying this whole issue are perhaps best illustrated by one specific incident.

In the spring of 1987 the editor was approached by members of Greenpeace Netherlands, who were intent on blocking the pipeline that discharges radioactive waste from the Sellafield Nuclear Reprocessing Plant in Cumbria, out into the Irish Sea. Those individuals wanted to be clear about possible charges and convictions that could be made against those taking part in the operation.

He advised them of the various criminal charges that were possible. He also explained that the operators of the Sellafield Plant, British Nuclear Fuels Ltd, would probably obtain an injunction and that, once it was served on them, they risked imprisonment if they breached the injunction. He suggested, however, that this was very unlikely and that the most probable outcome was that they and/or Greenpeace Netherlands would be heavily fined.

Some three months later the Greenpeace ship, the *Sirius*, anchored just off the shore adjacent to the Sellafield site, and activists started trying to block the pipeline. Meanwhile, British Nuclear Fuels' lawyers obtained injunctions from a judge in his bedclothes late at night ordering the recipients to discontinue the operation, and these injunctions were helicoptered back to Sellafield, where they were dumped onto the ship's deck. The injunctions were ignored and the pipe was blocked. At the resulting contempt proceedings the coordinator of the campaign, Hans Guyt, and the captain of the ship, Willem Beekman, were not only fined, as they had been advised, but the judge also decided to make an example of the pair and they were given a prison sentence of three months each.

The memory of visiting Hans Guyt in Pentonville Prison on New Year's Eve 1987 will long remain with the editor as a salutary lesson on the law's unpredictability in this field.

Pollution control

Chapter 11 describes how the process of pollution control operates in terms of the regulatory authorities. This is of vital importance in many cases where bringing problems to the attention of the authorities maybe the quickest and most efficient way of having a pollution problem resolved.

In some areas the regulatory bodies are effective at controlling pollution but it is too often the case that their pragmatism interferes with the interests of the environment, which has been all the more the case since the establishment of the Environment Agency in 1995. It is becoming increasingly apparent that this new body, formed to amalgamate Her Majesty's Inspectorate of Pollution, the National Rivers Authority (NRA) and the various waste authorities, has played down its enforcing role certainly in comparison with the more vigorous NRA.

Part III: Actions to claim from existing developments

There are three primary routes to claiming compensation from a company whose development is causing damage: claims for personal injury, claims for property damage and claims for nuisance. In addition it is possible to apply for injunctions in relation to the operation of a plant, where it can be shown that the plant is causing a nuisance.

Personal injury
Claims relating to the injuries suffered by individuals, considered in Chapter 12, are often complex because of the difficulty of proving that the injury was caused by the environmental pollution. Sometimes the link is obvious, as was the case with the victims of the Camelford affair, described in Chapter 12. In 1988, several tonnes of aluminium sulphide waste were dumped into the water system in the Camelford area, in error, causing an immediate attack of sickness among those who had drunk the water before the error was rectified.

What is much more difficult to prove is the existence of what are known as chronic symptoms – that is, complex illnesses such as cancers – which arise over a much longer period, and which are endemic to the population at large.

Using the Camelford example again, in that instance it was alleged that some of the Camelford residents, as a result of drinking the contaminated water, had suffered mental impairment. This claim was supported by the knowledge that aluminium is an accepted cause of such illness. However, there was enormous difficulty in showing that the mental state of those who had drunk the water was in any way different from what it would have been had they not drunk the water. Showing that somebody's IQ has deteriorated is extremely difficult. In the court action the difficulties were effectively insurmountable and the claimants concerned settled for amounts only marginally greater than the figures that related solely to the value of the sickness from which the residents had suffered.

Another example of the difficulties of proving that complex ailments have been caused by particular pollutants is the Sellafield childhood leukaemia case. In that instance the plaintiffs were trying to show that the excess of leukaemias and non-Hodgkins lymphomas around the Sellafield nuclear plant had been caused by radiation from the plant. The allegation was that either the radiation had mutated the sperm of the radiation workers who fathered the leukaemic children or alternatively that the radiation emanating from the plant had affected the children directly, bringing about the onset of the leukaemias.

There was undoubtedly a significant excess of childhood leukaemias in the vicinity of the Sellafield plant and it was accepted that radiation can cause leukaemia. Despite this there were enormous difficulties in proving that it was the radiation, rather than some other factor, that had brought about the excess. Although experts came from across the world to the High Court in London to suggest that this was the case, in the end British Nuclear Fuels and its lawyers were able to persuade the judge that the plaintiffs had not shown that 'on the balance of probabilities' it had been the radiation that had caused the childhood leukaemias.

The problem for those alleging that their illness has been caused by a particular pollutant is that scientists are still a long way from being able to determine the cause of most complex illnesses. There are very few illnesses that are specifically related to a particular pollutant. The best example of such an illness is mesothelioma, where the only known cause is exposure to asbestos. Where, therefore, an asbestos worker has contracted mesothelioma, the defendant companies have nearly always accepted that the mesothelioma was contracted as a result of the exposure of the worker to asbestos fibres.

Similarly, in the case of the two plaintiffs who lived within the immediate vicinity of the Turner and Newall plant in Armley, Leeds, the court held that there was such a strong correlation between the level of asbestos dust where they played as children and the mesotheliomas they contracted that it was satisfied that the two were causally connected. These are, however, the exceptions rather than the rule. When it comes to court decisions in this field the judiciary is reluctant to overcome the mental hurdle of linking environmental pollution to specific chronic illnesses.

Loss and damage
Companies involved in the production of environmental pollution seem to be rather more prepared to compensate when it comes to loss of property or financial loss than they are when it comes to injuries. It may well be that they are less worried about the publicity factor involved in such cases. Chapter 13 considers the issue of loss and damage, exploring the various ways in which damages can be claimed and examining those cases that have been more successful than others.

There are many examples of where people have claimed for financial loss in relation to environmental damage cases. The claims are usually, although certainly not exclusively, made under the common-law heads of nuisance, which is covered separately in Chapter 14, and negligence. Here are some of the claims that have been brought:

- Michael Saltmarsh, who owned Croyde Bay in Devon, successfully sued the regional water company for all the financial losses caused by the sewage that they pumped into the sea and that subsequently arrived back on his beach (see Chapter 15).

- Three farmers living in the vicinity of the Coalite plant in Derbyshire successfully obtained compensation from the operators of the plant when they were able to show that high levels of dioxins had emanated from the plant and contaminated their farms.

- The fishermen and businesses in the town of Tenby have largely been able to gain compensation following the spillage of oil from the *Sea Empress*.

- However, the Merlin family who lived down the coastline from the Sellafield nuclear plant, were unable to gain compensation for property devaluation, despite the fact that they were able to show that levels of plutonium in their house dust were higher than any other recorded levels in the western world – evidently the result of the radioactive waste pumped into the Irish Sea by the nuclear plant operators being brought back onto land by the tides (see Chapter 13).

- This decision can be contrasted with that in the case of the claim by Blue Circle Cement, who successfully sued the Ministry of Defence for the damage caused to its property through an accidental spillage of radioactive waste containing levels of plutonium far less than those found on the Merlins' land. If ever there was a comparison that suggests there is still one law for the rich and powerful and another for the ordinary person in the street then this was it.

- The residents of Armley have been unable to gain compensation, so far, from Turner and Newall for the diminution of the value of their properties caused by the perception of risk in living so close to the Turner and Newall factory and the perceived increased levels of asbestos within those homes.

Nuisance

Claims for nuisance are claims for the annoyance and/or inconvenience related to the setting up and/or operation of a plant. There are many examples of cases where the operation of some industrial process or another has been held by the courts to be a nuisance to the neighbours of the plant. In terms of winnability, these actions without question come top of the list. Chapter 14 examines in detail the circumstances where such a claim might well succeed.

Water and evidence

Finally, the book has chapters on the specific issue of claims relating to water (Chapter 15) and then the issue of what needs to be put before the courts by way of evidence if a case is to be won (Chapter 16).

The authors hope that this book will provide the necessary information to any individual or group considering using the law as a tool to prevent environmental damage or to obtain redress and reparation where such damage has already occurred.

Part I

PREPARING FOR ACTION

2   First Steps

Environmental legal actions usually commence when local people realise that something already happening – or, more often, something about to happen – is likely to impact adversely upon their lives. How does an environmental problem for a local community or individual citizen translate into setting up a group and taking legal action? This chapter looks at the first steps that are usually taken in this process.

The story of what happened to a small community in Wales is described below as a useful example of how local environmental concerns about the impact of pollution on health and the environment can be translated into legal action. Setting up a local group is usually a crucial step in this process and that issue is dealt with in some depth following the example.

CEFN MAWR

There has been a chemical works in the village of Cefn Mawr for decades. By the early 1990s the works were being run by the Sulphonamides Groups of Monsanto plc, a multinational chemical corporation. The plant employed about 500 people in a 24 hour, seven days a week, continuous production system. The company manufactured chemical additives that were used in the production of vulcanised rubber, essential for the production of tyres, Wellington boots and other rubber products.

The production process involved the use of a wide range of hazardous chemicals, including liquefied chlorine gas, liquefied ammonia gas, hydrogen sulphide gas, and a range of solvents such as toluene and trichloroethylene, acids and phenols.

The peculiar smells for which Cefn Mawr became renowned were largely the result of fugitive emissions of hydrogen sulphide and carbon disulphide, both of which produce a characteristic 'rotten egg' smell. At relatively low concentrations these chemicals are capable of causing nausea and headaches; high concentrations can cause problems to the eyes, skin and the respiratory system. Particularly high concentrations can, on occasion, be fatal.

The village, on the banks of the River Dee, to the south of Wrexham, was traditionally the source of much local humour. Cefn Mawr was the place where car drivers wound up their windows on travelling through to Llangollen; it was the place where the locals smelt rather strange. However, the people of Cefn Mawr were prepared to put up with both the smell and being the butt of the surrounding area's jokes because the works had become a local institution. Almost everyone's father worked there. The village children went to the works Christmas party and bingo sessions were held in the works social club. The local population accepted the fact that occasionally the air was not as sweet as it could be, and that their health did not seem to be as good as it could be, but then employment was high and so were wages.

Janet Williams, who had grown up in the village, returned to Cefn Mawr in 1986 and moved into a cottage close to the plant in 1991. Having lived away from the plant for a number of years her tolerance to the smell, built up during childhood, seemed to have diminished and both she and her partner, Andy Radford, began to suffer ill effects from the aerial pollution emanating from the plant.

Janet's first child, Jamie, was born in the spring of 1992. Jamie was born very underweight, he had breathing problems and he seemed to be persistently coughing and vomiting. He was eventually diagnosed with asthma at the age of six months and his parents suspected that this may have been caused by or exacerbated by the pollution from the plant. These concerns were heightened when, at the local Health Clinic, Janet began to realise that there seemed to be a large number of children suffering from repeated infections, eczema, hyperactivity and a surprisingly high level of colds and flu.

The couple expressed their concerns to the company on a number of occasions but they were invariably reassured by their representatives that the emissions were harmless. Then, in August 1992, a vent bag burst at the plant, showering cyclohexylmercaptan into the air. Mercaptans are notoriously odorous, and are used in industry to trace gas leaks. A few parts per billion is usually enough to cause significant discomfort.

The smell caused by this burst was both intense and long lasting, leading to a number of village residents complaining to the local Environmental Health Department, the HMIP (Her Majesty's Inspectorate of Pollution, which has since been absorbed into the Environment Agency) and local councillors, and publishing a letter of complaint in the local paper.

Janet Williams and Andy Radford felt it was time to try to consolidate these individual concerns and they set up a public meeting to which about ten people came. Shortly afterwards a second and then a third meeting were held, by which time interest had begun to grow in the local community, leading to some 40 people attending. At the third meeting a group was formed, calling itself CARE, standing for Communities Appeal for Respect for the Environment. CARE's purpose was to support people suffering from the effects of pollution and to campaign for improvements to their environment.

In the early days the group received a very cool response from local residents and the local community council, with the prevailing attitude being that the works were vital for employment in the area and that any criticism might provoke the company to scale down operations or even to move elsewhere. As a result, those who became involved with the group felt ostracised and isolated from the rest of the community. Even friends and relatives found the subject of the works too difficult to bring up in conversation.

Notwithstanding these difficulties, the group continued to meet to determine the best ways of realising its objectives, and it was during one CARE meeting that it emerged that the mercaptan leak in August 1992 seemed to have had similar health effects on three families, and most particularly on their young children, who had begun to exhibit a range of symptoms. The children's parents told the group that a number of other families had been similarly affected and these were then contacted. The majority of the families did not want to become embroiled in any sort of action but some did, resulting in legal advice being taken on whether they had claims against Monsanto.

When the news that members of CARE were taking legal action against Monsanto filtered out into the community, those involved were put under enormous pressure to drop the action and stop the campaign for fear of what impact this might have on jobs.

By this time, however, the members of CARE had the bit between their teeth, and instead of dampening down their actions they decided to step up the campaign and adopt a higher profile. The group took the view that the attitude the community had towards them was largely the product of ignorance, lack of knowledge, and fear. They felt that if local people were better informed about the industrial chemicals and pollution that were emanating from the plant then they would better understand the concerns of CARE, concerns not just for the members of CARE but for the whole community.

The group decided to publish a newsletter, the CARE *Eco*, which would provide a means of communicating, in an accessible way, scientific information about industrial chemicals, pesticides, pollution and the environment. At the same time the newsletter provided a focus for establishing and networking with other local action groups with similar concerns.

The raising of the group's profile and the publication of the newsletter paid dividends. The attitude of local people began to change and representatives of the group were invited to attend Monsanto's Public Advisory Panel, which would have been unthinkable only a few months previously. There was still some opposition within the local community but as the group grew in size its respectability grew as well.

In November 1994 an incident occurred that was to transform CARE and the local community. Another major accident at the plant this time resulted in the release of hydrogen sulphide and carbon disulphide. A cloud of gas spread slowly through the village, affecting over 100 people. Emergency procedures were implemented and large numbers of people evacuated themselves from the village.

The size and scale of this new release, the number of people affected and the fact that it had come so soon after the formation of CARE all led to the group bringing in lawyers and to large numbers of local residents deciding to pursue claims for compensation against Monsanto. The attitude of the local community to CARE was transformed, almost overnight, as local people began to realise that it was possible to stand up against the plant and to demand action of the statutory bodies responsible for controlling pollution from the plant, provided that action was taken in a coordinated and concerted way.

In less than four years CARE had grown from a handful of concerned individuals to a group that had majority support in the local community. The local Environmental Health Department and local MPs now refer individuals with concerns about their local environment to CARE, and it has become one of the largest grassroots organisations in North Wales. In the words of one of the founding members, 'If you arm ordinary people with the facts and give them the courage to use them, they can change anything they desire, for change will only occur when the majority recognise the need for it.'

SETTING UP A LOCAL GROUP

The experience of the group at Cefn Mawr has been mirrored many, many times throughout the UK over the last 25 years. From these experiences it is possible to draw together a number of guidelines that might be of assistance for anyone considering setting up a local group to campaign on environmental issues at a local level. (More detailed information on local campaigning can be found in Ward 1992.)

Probably the most significant issue for any individual considering the formation of a campaign group is the need to be absolutely clear about the objectives of such a group. A simple statement (preferably no more than one or two sentences) encapsulating the purpose of the group will provide a useful focus for prospective members and will help give direction to the group's activities. It will also ensure that the public and media are clear about what the group is trying to achieve. The objectives of different local groups may be varied: for example, opposing industrial pollution, as in the case of Cefn Mawr; opposing proposed developments, whether of local roads or supermarkets on greenfield sites; campaigning for the restoration of land following the closure of industry, or contamination incidents.

Having defined the focus of the prospective group, steps should then be taken to investigate whether any other group with similar objectives exists within the area. A useful place to start is usually the local library, and the local council may also have information – the environmental health department or a nominated Local Agenda 21 officer would be useful initial contacts when seeking information. National or other local environmental groups may also be worth contacting. If other groups with a similar aim do exist within the area then it may be sensible to consider joining forces. In some situations forming a completely new group might be the best way of capturing local concerns, but in others using the experience of an older group might give those concerns a greater focus, and might provide a readily usable medium for the communication of ideas and the achievement of goals.

Even if there are no existing groups that could absorb the new concerns it is well worth going to the effort of making contact with other organisations that may be prepared to help in setting up the new group, with structures, contacts, suggested meeting places, potential support and so on. Some of the larger national environmental organisations may be prepared to make small donations towards the initial costs of setting up a local group and, certainly, networking in the environmental field is usually of great benefit to any local group. A list of the main organisations and their contact details is given at the back of the book (page 405–14).

The practicalities of organising a local group are relatively straightforward: finding two or three people who share the same aims and objectives to form the nucleus, amending the mission statement to ensure agreement, and establishing a preliminary division of labour. The search for other members will partly depend on whether a specific area is being targeted or whether a workplace, college or other institution will provide the focus for the group. As a general rule, the more specific the targeted group the easier it is to focus publicity and make arrangements for meetings.

Launch meeting

Most embryonic organisations move quite rapidly from being a loose, informal core group of people to establishing themselves in the public domain by means of a formal launch. A meeting formally launching the group has a number of advantages. It is likely to be covered by the local press (provided they have been given adequate notice); it lets local people know that the group actually exists; potential new members can attend and find out more about the group; and members themselves can meet and exchange ideas. In addition, holding the meeting will allow the group to assess the local reaction to the new organisation. Arranging and holding the meeting is also vital for the group dynamics, because an organisation that takes no action will quickly lose its membership, focus and interest.

The form of the public launch will largely depend on the purposes of the organisation. The launch may be needed to establish the group, to recruit volunteers, to raise money, to discuss the concerns of the group, or to put a warning shot across the bows of the polluter, and is likely to be a mixture of all these.

A date should be chosen some weeks in advance and, if possible, it should not clash with other competing events (whether local or national). Before the meeting a timetable should be agreed, setting out the various tasks.

Guest speakers
A speaker who has experience in the relevant area but is also well known can help to attract people and the media, and can lend credibility to a newly formed group. The meeting date should be agreed with all guest speakers from the outset and the speakers formally invited. The speakers' attendance should be confirmed four weeks before the meeting, and they should be reminded again a week or so before the meeting.

Venue
The venue should be accessible by both public and private transport and should, if possible, have access for the disabled. Crèche facilities should also be considered. Some thought should be given about the numbers likely to attend. Any proposed venue should be visited and checked for suitability before the booking is made. The booking should be confirmed four weeks before the meeting, and double-checked about a week before the meeting.

Equipment
Attention should be paid to details such as whether it will be possible to black out the venue for slides or videos, and to ensuring the mechanical equipment is in working order. Many a meeting falls into a slough when a key speaker is just about to show a video or set of slides that are crucial to the points being made, and the machinery breaks down.

Costs
The group should have a clear idea of the meeting's costs, including the cost of hiring the room, advertising the meeting, postage and the fees of any guest speakers. The source of the funds to meet these costs should also be clear, with estimates of likely donations by those attending the meeting being necessarily conservative. Having a shortfall in funds when there are large bills to pay for an expensive meeting room and a guest speaker from foreign parts can be too much of a burden for even a well-founded group at the outset of its life.

Publicity
A publicity leaflet and/or poster should be produced as soon as the booking is arranged. Four weeks before the meeting it should be circulated and displayed as widely as possible, with further steps being taken to organise the publicity for the meeting, such as sending a press release to local press, radio and television. The press in particular may have diaries of forthcoming events. Around a week before the meeting a membership leaflet should be prepared, or a sheet for those who are interested in joining the group to fill in at the meeting.

Invitations

Four weeks before the meeting, individual invitations may be sent to people who have already expressed an interest in the group. Two weeks later, various local groups with similar objectives, as mentioned above, can be invited, although some care should be exercised in the selection process. For example, there may well be people who have separate agendas who might want to use the meeting to push their own interests. Such people should not be invited and, if it is thought that they are still likely to attend, the choice of the person chairing the meeting will become very important.

The chair

It is natural for the person who has put most into the group to want to chair the meeting. That person may have the strength of character to take on that role but, if not, thought should be given to having a person who is well known in the community and who is sympathetic to the group's interests, to take the chair. Open debate will be encouraged by a sympathetic chair. If the group wants decisions to be made a tougher and more dominant chair may be more appropriate. It is vital that the chair is someone who can capture the mood of the meeting, and who has the strength of mind to allow the flow of the proceedings to be flexible while at the same time retaining control.

The agenda

In the final week before the meeting a timetable for the meeting itself should be agreed with the chair, leaving enough time for the main business on the agenda, as well as time for clearing up after the meeting and talking to people attending the meeting who may wish to know more.

The agenda should allow for introductions, for the main speaker's address, and for any questions to be taken. There also should be time for the chair to take further practical steps in setting up the group if he/she senses the meeting could make progress.

It is quite possible that the first meeting will need only to focus on preliminary concerns such as agreeing that there is a problem and that the time is ripe for action to be taken. It may be possible to agree on a name for the group (ideally one that's reducible to a suitable acronym) or it may be that people could be invited to come up with something better by the time of the second meeting. Depending on the strength of public feeling, the urgency of the situation and the sophistication of the group, the meeting may be able to take matters further, deciding on future action.

Follow-up meeting
A follow-up meeting may well be necessary to focus on particular practical issues and the date might sensibly be fixed at the initial meeting. The practical issues may include agreeing a name for the group, a programme of work, the aims of the group, and selecting officers, albeit informally. It is sensible to appoint someone in the role of treasurer at this point, to carry out the initial task of setting up the group's bank account (for which the group's name will be needed).

At an early stage the group should consider taking legal advice, which should initially be available free of charge. A lawyer may be invited to a follow-up meeting to explain the legal options open to the group and how they may be funded.

If anything, the follow-up meeting is probably more important than the first. It will be the first time that those wanting to be involved will have sat down as a group and discussed practical issues. It must be well planned and the chair should pay particular attention to what needs to be achieved. Agreement on basic structures at an early stage will reduce bureaucracy to a minimum.

Appointing officers
Most local, voluntary, groups have a tendency to collapse after a period, even if the purposes of the group have not been achieved. One of the most common reasons for this is that the coordinator or person with the most energy leaves the group and no one else is prepared to devote as much time and energy to the project. For this reason, it is essential that, from the start, work is shared among group members as widely as possible and the structure for the group is capable of accommodating change, growth and development as the group evolves.

Groups tend to develop similar structures in the course of time. The coordinator of the group is usually the driving force behind the group, may well chair meetings, and will tend to be the public face of the group dealing with press and media. Second to this person is usually the secretary, who is responsible for administration of the group. The separation of the role of these two posts needs to be clear early on to avoid duplication and treading on toes.

The treasurer is also essential to any group, as even the smallest of groups will be involved in financial transactions of some sort. If proper accounts are not kept the potential for disagreements over money within the group – and even for mishandling of the group's money – increases substantially. The other key financial position is that of fund-raiser, who must feel comfortable about approaching the public for funds. A membership secretary should also be appointed.

Although the coordinator of the group is likely to have most contact with the media, an individual may be delegated to deal with all publicity, including recruiting volunteers, producing information bulletins or newsletters, drafting press releases and so on. If a lawyer is instructed, the group of individuals instructing the lawyer will need to decide what relationship they want that lawyer to develop with the media.

PUBLICITY AND MEDIA

Effective publicity and the use of media can make or break an environmental group and this is discussed more fully in Chapter 4. Publicity is essential for:

- *recruiting new supporters* – encouraging other people to become involved in the campaign and letting wider numbers of people know of the group's existence;

- *fund-raising* – generating an income, attracting support from people who are prepared to give money rather than time;

- *raising awareness* – letting people know the group's aims and showing how necessary the group is;

- *pressurising the polluters* – frightening the polluting company into resolving the problem in order to protect its image.

Most groups, once established, produce a newsletter of sorts to keep members and the media informed of developments. Posters are invaluable for publicising particular events, and the more established groups have branched out into producing other merchandise, such as T-shirts and car stickers, to put the group on the map. The group may wish to produce a campaign video. If so, *The Video Activist Handbook* (Harding 1997) provides relevant guidance.

Developing a working relationship with the media is of paramount importance for environmental action groups. At a local level, the group should work closely with its regional newspapers, radio and television stations. Although television coverage can be difficult to achieve, if issues are presented in a way that appeals to producers then it may be possible for particular events to be highlighted. Environmental issues remain high on the public agenda so there is a fair chance that local and possibly even national media will show interest in these types of campaign.

A well-drafted, clear press release should attract the attention of television, radio and the press. A press release should always state what is happening, where it is happening, why it is happening and who is involved. It should contain a quote from a member of the group that can be used in reports, and should always contain contact numbers where journalists can obtain some background to the story, which may include contact details for any appointed lawyer to update the media on the legal position. It goes without saying that ample notice of developments should be given to the media, whenever possible.

The more sophisticated group coordinators will consider granting exclusives and imposing embargoes on media information. However, these can be more trouble than they are worth. They can also be counterproductive and are often difficult to impose, particularly if groups are not well established and have little leverage over the media, so they are generally best avoided.

CONCLUSION

Citizens facing a new development or problems with an existing one should not be overly daunted by the prospect of fighting the change. Help is readily to hand through existing organisations, and by choosing experienced lawyers it is often possible to find a way to attack the proposals effectively. That is not to say that such attacks are always successful, but there are many examples of developments being stopped in their tracks by concerted opposition. The quality of our environment is too important to be left solely to the politicians and the regulators. It is only through the vigilance of local citizens that we have a real hope of protecting our environment for our children and grandchildren.

3   Funding

When individuals who are suffering some sort of damage or loss from environmental pollution turn to the legal profession for an assessment of their rights and the possibility of pursuing a legal action, at the top of the agenda will be the issue of how such an action can be funded, whether in the courts or through the presentation of a case to a public inquiry or other tribunal.

Taking legal advice and considering legal action is often regarded as a last resort by the public in general because of the significant financial implications involved in being a party to legal action. This chapter considers how cases can be funded.

The role of obtaining legal aid belongs primarily to the lawyer. However, more people nowadays are applying for legal aid themselves and then, with the benefit of a certificate, appointing lawyers. Although some lawyers undertake this task on behalf of their clients, it was felt that it would be helpful to provide a degree of detail, both to assist those who are applying for themselves and to clarify the process for those for whom applications are being made by lawyers.

GENERAL RULES ABOUT COSTS

It is useful to start by considering the general rules that apply in relation to costs when embarking on a legal action. There are two separate funding issues that an individual has to face when considering whether or not to bring a legal action.

The first is that of how the costs of the proposed action are to be funded – for example, the costs of employing a solicitor,

barrister, and medical and/or technical experts. This issue is all
the more important because cases involving environmental
pollution and associated personal injuries are generally expensive to run because of the level of technical information that
such cases require.

The second issue relates to the application of the general rule in
litigation that the loser pays the winner's costs. Where an
individual fails, for example, in a claim for nuisance against a
noisy and smelly factory, the factory owner will obtain an order
that his/her legal costs must be paid by the individual concerned.
Therefore, where a case is lost there is the potential for an
individual to be liable for both sides' costs. The converse is also
true: that is, where a plaintiff's claim is successful, that person will
benefit from the 'loser pays' rule and the majority of the costs will
be paid in addition to any damages claimed.

It is these two factors that make funding issues in environmental litigation of vital importance. The purpose of this chapter is to
explore the methods of funding this type of litigation or other
forms of legal representation. Because of the scale of costs in these
actions and the prospect of also having to pay the defendants'
costs if the action is lost, the vast majority of environmental
claims have, to date, been funded through the legal aid scheme,
which is described in detail below. However, legal aid funding has
been under ferocious scrutiny in recent years and it may well be
that, whichever government is in power in years to come, the
ability to gain legal aid will become more and more difficult. With
conditional fees now being allowed for personal injury claims
(described below, page 57–61), it may well be the case that state
funding in this type of action will be withdrawn altogether.

LEGAL AID

Throughout the UK there is a state-funded legal aid system which
provides legal advice and assistance to those people who are able
to satisfy the eligibility criteria, whether they are acting as
plaintiffs or defendants in civil actions, or as defendants in
criminal cases.

Legal aid is available for criminal cases and for certain types of civil actions. Criminal legal aid is available to those who face serious criminal charges. For example, an environmental activist who has taken part in a demonstration may well need legal advice and representation under the criminal legal aid system if criminal charges are brought in relation to offences such as criminal damage or trespass. Civil legal aid is available to assist in the bringing of claims for damages for personal injuries, losses or nuisance, such as those caused by noxious emissions from an industrial plant or factory, and for the judicial review of decisions made by public authorities. It is, however, not available for planning inquiries or for inquests.

The tests that determine whether or not an individual is eligible for legal aid differ, depending on whether a person is applying for criminal or civil legal aid.

THE CRIMINAL LEGAL AID SYSTEM

The Duty Solicitor Scheme

Where an individual has been arrested and taken to a police station for questioning, advice can be obtained under the Duty Solicitor Scheme. This scheme was established, following a review of the existing court duty solicitor scheme under the Legal Aid Act 1982 and the Legal Aid (Duty Solicitor) Scheme 1983, to establish police station cover. It came into force on 1 January 1986, and provisions relating to it were included in the Police and Criminal Evidence Act 1984 (PACE 1984).

Advice under this scheme is available to everyone irrespective of their means and is non-contributory. The scheme pays for solicitors to attend at the police station at all times of the day or night, to advise anyone who has been arrested, and who requires the assistance of the duty solicitor, about their rights, and to be present during police questioning.

Having a solicitor of choice in any circumstance is undoubtedly far better than having an unknown quantity like the duty solicitor. However, at the same time, having the advice of a solicitor is almost always preferable to acting without any assistance.

Under PACE 1984 and the Codes of Practice made pursuant to that Act, a custody officer must give the person detained at the police station information regarding the right to contract privately with a solicitor (either one of his/her own choice, or the duty solicitor, or one chosen from a list held at the police station) and that independent legal advice is available free of charge. If the person elects for a consultation with a duty solicitor, the nearest duty solicitor available in the district will be called to give advice either by telephone (for very straightforward cases) or in person, by attending at the police station.

The duty solicitor scheme is available to anyone who is:

- arrested and held in custody at the police station or elsewhere;

- being interviewed in connection with a serious offence under armed services legislation;

- attending the police station voluntarily to assist with police investigations, which may include victims of crime.

The initial financial limit allows two hours of legal advice, which includes time spent advising and assisting on the phone and in person plus travel and waiting time. It is possible for the solicitor to extend this limit retrospectively where assistance is required in the interests of justice and has been given as a matter of urgency.

Administration

The criminal legal aid system is administered by the criminal courts. Where a person is charged, the application for criminal legal aid is made to the magistrates' courts. In order to obtain criminal legal aid it is necessary to show officers of that court that the accused satisfies the means and merits tests set by Parliament.

The means test looks at the applicant's disposable income and capital. Where an individual is outside the financial eligibility limits the case will have to be dealt with on a private basis.

The court has a wide discretion as to whether or not to grant legal aid and will do so where it considers that this is desirable, in the interests of justice. The factors that the court will take into account when deciding whether to grant criminal legal aid are set out under section 22(2) of the Legal Aid Act 1988. Some of the most important considerations are whether:

(a) the accused is likely to be deprived of liberty or to suffer serious loss of livelihood or serious damage to reputation as a result of conviction;
(b) the case may involve a substantial question of law;
(c) the accused may not be able to understand the proceedings or state his/her case by reason of inadequate English or mental or physical incapacity;
(d) the case involves the tracing and interviewing of witnesses or expert cross-examination of prosecution witnesses;
(e) it is in the interests of someone other than the accused that the accused should be represented.

The magistrates' courts generally apply the Justices' Clerks Society Guidelines. These guidelines have no legal status, but they indicate that legal aid should:

• always be granted for offences triable only on indictment (i.e. those capable of being tried in the Crown Court);

• normally be granted for certain offences, such as affray, actual bodily harm, burglary, criminal damage;

• not normally be granted for Road Traffic Act offences, drink charges, prostitution charges, fare evasion and litter offences.

In the magistrates' court, legal aid is normally granted for representation by a solicitor only. A barrister may be authorised where the defendant is charged with an offence that is triable in the Crown Court, before a jury, and where the circumstances are considered unusually grave or difficult so that representation by both solicitor and barrister is desirable. In all other courts, and normally in the Crown Court, the legal aid order provides for representation by solicitor and barrister.

If legal aid is granted it will allow an individual to be represented by a solicitor who will prepare the case, analyse the prosecution evidence and find witnesses to support the defence. The solicitor will then advise the person charged about the law and the sentence they face if found guilty of the charges brought. If the charges are serious, the legal aid system will also pay for a barrister to advise and represent an individual at trial. It will also pay for any technical experts needed to advise or assist in the gathering or analysis of the evidence.

If legal aid is not available, either because the charges do not pass the 'interests of justice' test or an individual is outside the financial eligibility levels, it is possible to hire a solicitor privately. Under the court duty solicitor scheme (which is run similarly to the police station duty solicitor scheme) the duty solicitors at the magistrates' court are available to give advice to people who have no representation.

Where an individual has been arrested as a result of a concerted campaign by an environmental pressure group, it is often the case that the group will have organised for lawyers to be available, who will be called upon where arrests are made for advice and representation where necessary. If not, sympathetic lawyers can often be found to give *pro bono* (i.e. free) advice where this is required and legal aid is not available.

THE CIVIL LEGAL AID SYSTEM

Administration

The civil legal aid system is administered by the Legal Aid Board. The Lord Chancellor's Department has overall control of the legal aid system. The Board is organised into various regional offices which administer the legal aid system in that region and deal with all civil legal aid applications from solicitors operating within their boundaries.

When approaching a solicitor for advice, it is important to find out whether he/she undertakes legal aid work, as this is by no means always the case.

The Legal Aid Board has recently set up a scheme whereby it

awards 'franchises' to firms of solicitors in certain areas of work if they are able to demonstrate compliance with quality assurance standards set by the Board. It is possible, by contacting the local legal aid office, to obtain a list of firms that are franchised in certain areas of work – for example, personal injury and crime. There is no franchise as yet specifically for environmental claims.

The civil legal aid system provides assistance to individuals seeking to bring or defend civil claims. In environmental litigation cases, they are far more likely to be applying for legal aid to bring rather than defend claims. Irrespective of whether the individual is an intended plaintiff or defendant, the same tests are used for determining whether or not legal aid should be granted in civil cases.

Initial advice

People who are on very low incomes or in receipt of income support may be eligible for free initial advice or advice subject to the payment of a modest contribution under the Green Form advice scheme. This scheme allows an individual two hours of legal advice from a solicitor, who can apply for extensions where this period is insufficient to deal with the problem.

The Green Form scheme usually allows for the solicitor to interview the client, find out the nature of the problem and advise the client as to the legal rights and remedies available. If a full grant of civil legal aid is available, the solicitor will usually complete the legal aid forms and submit an application for legal aid to the Legal Aid Board following the initial interview with the client.

As has been previously stated, certain types of advice are expressly excluded from the civil legal aid system, with the general rule being that civil legal aid will only be granted for cases that involve litigation between parties – that is, where legal proceedings have been or will be issued. This means that legal aid is not available for advice and representation at public inquiries, tribunals or inquests. (However, the Lands Tribunal, where land compensation claims under the Land Compensation Act 1973 are decided, is an exception to this rule because it is regarded as a subdivision of the High Court.)

Where full legal aid is not available, it may be possible to obtain extensions to the Green Form scheme to allow certain preparatory work to be done. However, it is not possible to obtain actual representation at a public inquiry under the Green Form scheme.

That said, although legal aid is not available for representation at an inquiry, if, after the decision has been made, a person wishes to challenge the decision on the basis that it was unlawful or that the decision-making process was unfair, then legal aid is available to bring a claim, known as a 'judicial review' (see Chapter 9) to the High Court to challenge the decision.

There are many people who are not eligible for advice under the Green Form scheme because the financial eligibility levels are set very low. If this is the case, some solicitors will be prepared to offer a free initial advice session. This will often include obtaining the facts of the case and advising on the relevant law and the remedies available. Since the financial eligibility levels for full legal aid are more generous than under the Green Form scheme, initial advice will often include the solicitor completing and submitting application forms for the full legal aid if this is available to the client.

How to apply

In a case that involves litigation between parties and where a solicitor considers that there is a sufficiently strong case to warrant public funding and that the applicant will be financially eligible for legal aid, an application for full legal aid will usually be made.

Civil legal aid is applied for by completing a number of forms. One of the forms asks the applicant to provide information about the case and state why legal aid should be granted. It requires the applicant or the solicitor to set out the basic details of the case: for example, the nature of the claim, its value, the facts, evidence and law on which the applicant will rely to support the claim.

The solicitor will usually but not always complete this application form after the first interview with the client. The solicitor will need to know all the facts that are relevant to the claim, and sometimes, where the case is not straightforward, the applicable law will need to be researched.

The second form that must be completed relates to the applicant's financial circumstances. Different types of forms are to be completed, depending on whether or not the applicant is in receipt of benefits (primarily income support). The applicant usually completes the financial form with the assistance of the solicitor.

LEGAL AID: THE THREE-STAGE ELIGIBILITY TEST

There is a three-stage test to decide whether civil legal aid should be granted.

1 The means test

The means test assesses whether an individual is financially eligible for legal aid. Broadly speaking, tighter financial requirements exist to assess whether a person is eligible for civil as against criminal legal aid.

Around 70 per cent of households in England and Wales were eligible for legal aid in 1984, but this had dropped to under 50 per cent in 1996. This was primarily achieved through the eligibility levels being specifically reduced by the government in 1993 and by a general failure to keep up the levels with inflation.

Individuals claiming income support pass the means test automatically and will be eligible for full legal aid without having to make any sort of financial contribution. This is also the case for children, unless they have a private source of income or capital, or they are in receipt of certain invalidity benefits. For those who are either working or perhaps claiming benefits other than income support, it is necessary to assess the income coming into the household, then subtract the amount of money spent on certain items such as housing costs, travel expenses to work and the number of dependants being supported in the household. This means that a calculation of income less outgoings must be made in order to assess the disposable income of the individual applying for legal aid.

Capital also needs to be considered carefully, as the Board takes into account not just savings but also other assets such the value of any jewellery and other valuables.

Generally speaking, working individuals in the middle-income bracket will usually only be eligible for legal aid where they have a number of dependants to support and high housing costs, such as mortgage repayments. Pensioners and those in receipt of family credit may often be eligible for legal aid, but subject to payment of a contribution to their legal aid. Contributions from income are paid monthly to the Legal Aid Board throughout the duration of the legal aid certificate, which is usually until the end of the case.

It is possible to assess whether a person is financially eligible for legal aid by obtaining and completing an up-to-date version of the Legal Aid Board's 'ready reckoner'. This can be found in the Legal Aid Board's leaflet *A Protocol Guide to Legal Aid*, pages 21–4. The financial eligibility part of the legal aid application is currently administered by the Legal Aid Section of the Department of Social Security, in Preston, although formal parliamentary responsibility for this function was taken over by the Legal Aid Board on 1 April 1997. Actual functioning will be phased in from 6 October 1997, beginning with four area offices, and the final phase is due to be completed by 1 February 1998. The Preston office is due to close at the end of March 1998.

2 The merits test

The second test for civil legal aid is the legal merits test. This is set out in section 15(3) of the Legal Aid Act 1988. The section states that a person shall not be granted legal aid for the purpose of any proceedings unless the Board is satisfied that there are reasonable grounds for taking, defending or being a party to the proceedings. This is known as the legal merits test.

In addition to the legal merits test, it is necessary to satisfy a reasonableness test, which relates to an overall concept of reasonableness, and which allows the Board to assess whether these are the types of proceedings that should be funded by the public purse. Therefore, to grant legal aid, the Board must be satisfied on the facts and the applicable law that there is a case to be made out or a defence that can be put before a court for a decision and that stands a reasonable prospect of success.

An application for legal aid is first considered by the Board's officers, usually based at the local office. Although it is not their role to review the case to the level of scrutiny accorded by the judge in deciding whether to grant legal aid, they will take notice of the availability of evidence to support the claim and assess the likelihood of success.

There is a box on the legal aid application form that must be completed, in which the applicant, or more usually the solicitor, is required to give an assessment of the chances of the claim being successful. Those completing the form are also required to address the question of whether, if the case is successful, the assisted person would be able to enforce a judgment against the defendant.

The Legal Aid Board will only fund litigation where there is a reasonable prospect that enforcement proceedings would be successful. Therefore, it will be necessary to show the Board that the defendant is insured in respect of the claim or has assets against which a judgment could be enforced.

The question of enforceability might be relevant to an environmental claim where, for example, the defendant is a company that no longer exists. It is certainly possible that in a claim for loss of property value, where a house was built on land that was contaminated decades before, it might be difficult to locate the defendant company and/or its insurers. In that type of case it would be necessary to provide the Legal Aid Board with some information to indicate against whom judgment would be satisfied in the event that the claim was successful, or what steps could be taken to investigate this issue.

In deciding whether to grant legal aid on the basis of the merits test, the Legal Aid Board must, without being over-cautious, avoid granting legal aid cases where there is little or no hope of success.

Generally speaking, there are reasonable grounds to grant legal aid if:

(a) there is an issue of fact of law that should be submitted to the court for a decision – for example, where the plaintiff says he was injured as a result of the defendant's negligent driving, but the defendant says that the injury was pre-existing before the accident;

(b) the solicitor would advise the applicant to take or defend the legal proceedings privately – that is, on the basis that the hypothetical person had adequate means to meet the likely costs of the case or would make payment towards its potential costs, although this would be something of a sacrifice;

(c) the application shows that, as a matter of law, there are reasonable grounds for taking or defending proceedings – that is, that there is a case or defence that has reasonable prospects of success, assuming the facts are proved.

To see how the system works in practice let us consider the example of a young construction worker who was employed for a number of years, without any protective clothing or respiratory equipment, on a site known to be contaminated with chemical waste. He developed a rare form of cancer known to be associated with the type of chemicals to which he was exposed. He was unable to work and support his family and his life expectancy was greatly reduced.

In that case, there were two questions that had to be answered. Was the applicant exposed to dangerous chemicals? If so, did they cause his condition or did it occur naturally? There was some *prima facie* evidence (that is, superficially strong evidence) that there was a link between the chemicals and the disease; there was also clear evidence of exposure and an indication that his employers might have been negligent.

Because of the seriousness of the condition and its financial implications for the client and his family, it was possible to say that a person of moderate means would be advised to investigate and possibly pursue a claim. There was at least an indication on the available evidence that there was a reasonable prospect that the claim would be successful.

However, it is often the case with environmental claims that it will not be possible to assess the prospects of success without first conducting lengthy investigations. The Legal Aid Board recognises this in some cases and will allow legal aid to be granted to allow investigative work to be carried out. However, it is still a requirement that there must be some evidence on which to base the assertion of a claim, when the application for funding is first made. In the example quoted above, if the client had developed a rare form of cancer but there had been no evidence of direct exposure to chemicals or agents that might have caused the condition, it is unlikely that legal aid would have been granted.

3 The reasonableness test

Having passed the merits test, an applicant must also be able to satisfy the reasonableness test. This allows the Legal Aid Board to consider, in all the circumstances, whether legal aid should be granted. Therefore, if after considering all the circumstances the Legal Aid Board decides that it would be unreasonable to refuse legal aid, it may grant legal aid. Although the Board is given a wide discretion under this test in deciding whether or not legal aid should be granted, there are well-established circumstances under which legal aid will be refused or granted, some of which are discussed in more detail below.

The value of the claim

One very important factor relates to the value of the claim being made. In all cases except those where a claim for personal injuries is being made, it is rare for legal aid to be granted where the financial value of the claim does not exceed £3,000. This is because costs will not usually be awarded to be paid by the loser in a case worth less than £3,000. Such cases are usually referred to arbitration in the county courts, where costs are only awarded in exceptional circumstances.

Since the cost of employing a lawyer could easily reach £3,000 in a contested action, it is deemed uneconomic for such low-value cases to be granted legal aid (see also the 'legal aid statutory charge', page 54).

For cases involving an element of personal injury, there is a lower financial limit of £1,000 for the award of costs to follow. This is because the Legal Aid Board and the courts regard personal injury cases as often being difficult to prove, and recognise that they will frequently involve more complex issues of fact, evidence and law.

Cost–benefit analysis

Another circumstance where legal aid will be refused under the reasonableness test is where the proceedings are judged as being unlikely to be cost-effective that is, where the benefit to be gained by the individual applying for legal aid does not justify the costs that will be incurred. For example, in a nuisance claim that is perhaps affecting a whole street of residents but where only one person wishes to pursue an action against the polluter, the Legal Aid Board may deem the costs of bringing an action on behalf of one person as being excessive when compared with the limited financial gain to that individual.

If, however, a significant number of affected residents wish to pursue the claim and accordingly apply for legal aid, the costs of the action would be spread among that group and the potential pool of damages would be multiplied by the number of people who wish to proceed. Therefore, it may often be the case that an individual claim for environmental nuisance will fail the reasonableness test because the costs of proceeding outweigh the benefits to be gained, whereas if a number of people get together to bring the action, the reasonableness test might well be satisfied.

Take the example of a factory fire that resulted in the evacuation of everyone within half a mile radius, and where all the evacuees suffered minor smoke inhalation, anxiety and inconvenience for one night. It would be uneconomic for one or two people to bring a claim against the factory. Their claim might be in the region of £800 each, but to prove negligence and bring an action would cost tens of thousands of pounds. However, if 200 of the evacuees claimed, this would make the total value of the claim worth potentially £160,000 and therefore, from the Legal Aid Board's point of view, it would be cost-effective to grant legal aid.

The financial gain from the claim, in the form of damages, is often of limited importance to the individual whose primary purpose in bringing an environmental claim is to stop the problem occurring in the future. This is particularly the case where an individual or group is trying to stop a nuisance that is affecting their community. This fact will be taken into account when assessing whether the reasonableness test is satisfied.

Are others likely to benefit from the claim?
In legal claims that involve environmental issues it is often the case that people other than those who have applied for legal aid to bring the claim will benefit indirectly from the litigation. Using the example above, where an individual wishes to claim damages for continual nuisance from a noisy and smelly factory and as part of the proceedings asks the court to issue an injunction against the factory to prevent future nuisance, other people in the area similarly affected by the nuisance will benefit from the legal action.

Understandably, the Legal Aid Board is concerned that people who are not eligible for legal aid may 'ride on the back' of a legally aided person's certificate and thereby benefit from the legal action to which they have not contributed. To deal with this problem, Regulation 32 of the Civil Legal Aid (General) Regulations 1989 gives the Legal Aid Board a discretion to take into account the existence of other people with the same interest in the action as those applying for legal aid.

Using the example of the evacuees above, if only 50 of the 200 people evacuated wanted to claim (perhaps the other 150 are not eligible for legal aid), the Board would probably consider that those who were not claiming could indirectly benefit from the grant of legal aid to the 50. In practice this is a difficult question. It is not yet clear how far the Board will apply the 'same interest' test to cases where people are applying for legal aid to protect themselves from environmental pollution. There have been no legal cases asking the court to interpret what the 'same interest' means in these circumstances.

It is possible that this test applies only to cases where people have a contractual or statutory relationship (for example, tenants in a block of flats), rather than in cases where people are claiming for personal injury or nuisance, where, inevitably, each person's interest in the litigation will be slightly different. Therefore, it can be argued in environmental claims, where a number of people are affected by the problem but they do not all intend to join in the legal action, that it is unreasonable for the Board to seek a contribution from the other people affected. They do not have the 'same interest'.

It may also be possible to circumvent this problem by providing evidence that specifies the different ways in which the person or persons applying for legal aid are particularly affected by the environmental pollution.

Can other bodies deal with the problem?

Another reason why legal aid is often refused in environmental pollution cases is because the Board considers that the application for legal aid is premature and that there are other bodies or organisations that can take action to prevent the problem. This is a common basis for refusal where the legal aid application relates to injunction proceedings to prevent a nuisance occurring. Since it is often the case that the individual's primary objective in bringing legal action in relation to nuisance will be to stop it, rejection on this ground can be a serious obstacle.

The main bodies contemplated by the Legal Aid Board as being able to take action in relation to an environmental pollution problem will be the local authority, the Environment Agency – previously Her Majesty's Inspectorate of Pollution (HMIP) and the National Rivers Authority (NRA). In reality, however, because of limited resources, it is often the case that local authorities and other enforcement agencies are not prepared to take enforcement action. The Board usually accepts this and will grant legal aid in such circumstances.

It is helpful when applying for legal aid if individuals can show that they have asked the enforcement agencies to take action in relation to the problem but that nothing has been done or the response to the problem has been inadequate.

Overriding importance to the client

Even if an individual fails to meet the standard for the cost–benefit test (that is the costs are likely to outweigh the benefits to the plaintiff), the regulations allow the Board to grant legal aid where the case is of the utmost importance to the client. (Surprisingly, there is no equivalent provision relating to the 'importance to the public in general'.) For example, where parents consider that one of their children has died or has suffered a very serious illness as a result of environmental pollution, such as radiation exposure, it may be possible for them to obtain legal aid on the grounds that the case is of utmost importance to them.

The primary test for the Board in applying this provision is whether a person of reasonable means, paying privately for the case, would pursue the action. This provision is rarely relied upon, but is useful in certain special cases that might not strictly satisfy the tests required for the grant of legal aid. However, there is no discretion available if the person applying for legal aid fails to meet the financial eligibility test.

An example of where this exception to the general rule was used was that of Simon Studholme (deceased). In that case it was alleged that Simon's death, resulting from the onset of leukaemia, had been caused by his exposure to high levels of electromagnetic fields. The value of the case was assessed at around £100,000, with the costs assessed at well in excess of that figure. The Board officers refused the application to extend the legal aid because of the failure to meet the cost–benefit test, but this was overturned on appeal on the basis that the determination of whether Simon's death had been caused by his having slept with his head effectively against the electricity meter in the house was of the utmost importance to the family.

How civil legal aid applications are processed

When the legal aid forms have been completed, the solicitor will usually send the forms to their local legal aid area office, with a covering letter setting out the basis of the proposed action. In all cases, an officer of the Legal Aid Board will consider the application at the outset. If the officer considers that the merits and reasonableness tests are satisfied, then the legal aid certificate will be issued where the person is financially eligible for free legal aid.

Where a person is eligible for legal aid but subject to the payment of a contribution, the Legal Aid Board will send out an 'offer' of legal aid to the applicant. In order to accept an offer of legal aid, the applicant needs to send the offer back with the first instalment of his/her contribution.

If the officer of the Legal Aid Board considers that the case does not satisfy the merits and reasonableness tests, then legal aid will be refused. A letter setting out the grounds of refusal will be sent to the solicitor and the applicant, and there is an automatic right of appeal from this refusal. An appeal against the refusal of legal aid will be considered by a committee of lawyers known as the 'area committee'.

Each of the regions of the Board has a large pool of barristers and solicitors from the region, who are members of the area committee, and they sit every few weeks in teams usually of three or four to hear legal aid appeals. They will decide the legal aid application afresh.

It is possible to make oral representations to the area committee and, in addition, to make further written representations to the committee, for example, to answer the grounds of the refusal. A solicitor will usually help the assisted person at the appeal and will often go, particularly in complex environmental actions, to make oral representations to the area committee.

Refusal of civil legal aid by the area office

Where the area committee refuses to reverse a legal aid refusal there is no further appeal structure within the legal aid system, but it is possible to challenge the decision in the courts in certain circumstances. This is done by bringing proceedings known as a 'judicial review' of the Legal Aid Board's decision (see Chapter 9 for more detail about this procedure).

It is not possible to bring a judicial review simply because an

individual disagrees with the Board's decision; there must be
something inherently wrong with the decision itself or the
decision-making process. For example, an individual might claim
that the decision was unfair because they were not given a fair or
impartial hearing, or that the decision was perverse – in other
words, that it could be shown that no reasonable area committee
could have reached that decision on the facts before it.

Surprisingly, perhaps, legal aid is available to challenge the
decision of the Legal Aid Board! A new legal aid application has
to be made to take this action, which is made to a different legal
aid area office.

The lawyers acting for an individual will need to consider
carefully whether or not they have grounds to proceed with a
judicial review of the Legal Aid Board's decision if legal aid is
refused. If successful, the court is not in a position to make the
decision to grant legal aid itself but must refer the matter back to
the area committee for a reconsideration.

The court will, however, usually order that a different group of
lawyers from the area committee – or even a new legal and area
office – decide the returned application. There are no guarantees
even after a successful judicial review claim that legal aid will be
granted, although the court may give an indication to the
applicant and the Legal Aid Board as to how the decision should
be determined.

Defendants' representations

It has become commonplace in the last few years for the
defendants in proposed litigation, particularly in high-profile
cases of public importance, to make representations when a legal
aid application is made. Many companies are extremely sensitive
to claims that they are affecting the environment or the health of a
community living close to their factory or operations. An individual bringing proceedings with the benefit of legal aid and
claiming that harm is being done to the local environment or
population could be a major blow to the company's public
relations. Defendants will, therefore, often employ solicitors

before legal aid is granted to make representations to the Legal Aid Board setting out their position in relation to the claim and attacking the basis of the applicant's claim.

As the defendants usually make their representations having heard of the potential claims through the media, there is something to be said for avoiding media interest in the potential claim until after legal aid is granted. There may, however, be cases for which generating publicity would be advantageous (see Chapter 4). The best course to take will depend on the circumstances of each case.

The Board will take the defendants' representations into account, but is obliged, through the interests of natural justice, to provide the applicants and their lawyers with a copy of the representations to enable them to respond. In high-profile environmental claims, the application for legal aid often becomes a battle with the defendants and the Legal Aid Board, ending with the appeal hearing becoming a virtual trial.

Lawyers acting for individuals in these circumstances will need to be prepared to respond to the defendant's representations, which will often be made in great detail and with the benefit of technical as well as legal advice on the allegations that the individual is making.

As an extreme example, in an application for legal aid on behalf of people who alleged that they had suffered illness from smoking cigarettes, the tobacco companies against whom legal action was proposed made five very lengthy sets of representations to the Board, the applications took some four years to determine, and there were four appeal hearings and two sets of judicial review proceedings against the Board's decisions. Even after all that the legal aid applications were refused.

What civil legal aid covers

Legal aid will cover all the work that the solicitor considers necessary to prove the case, subject to any specific limitation on the certificate, whether financial or in terms of the scope of the work allowed. It will cover the cost of the solicitor's time, pay for barristers' fees and allow a solicitor to instruct medical and technical experts and pay for court fees.

For example, in a claim for nuisance from the operations of a waste-transfer station, once legal aid has been granted the solicitor might take the following steps under the legal aid certificate:

1   Take statements from the clients.

2   Instruct noise and dust deposition experts to advise on the operation.

3   Having obtained these reports, instruct a barrister to consider the evidence and law prior to deciding whether legal proceedings should be issued.

Role of Legal Aid Board after granting civil legal aid

It is often the case that, in environmental pollution claims or claims that involve complex issues of causation, the Legal Aid Board, having granted legal aid, will keep a tight rein on the proceedings in terms of what the lawyer can do and the costs that can be incurred. The Board can do this by granting certificates that allow only certain specified steps to be taken.

Once those steps have been completed, the lawyer is required to report back to the Board, giving an assessment of the case at that stage. The Board then decides whether or not the scope of the certificate should be extended to allow further work to be undertaken. So, in the nuisance example above, it is likely that the certificate will initially be limited to investigation; then, after obtaining a barrister's opinion or submitting a solicitor's report, the certificate would be extended to allow proceedings to be issued.

There are two different types of limitations to a legal aid certificate. The first relates to the actual legal steps that can be taken. In a environmental action related to personal injury, these might initially include obtaining the applicant's medical records and a report from a doctor specialising in the relevant illness, in order to assess the nature and possible causation of the illness, addressing in particular the applicant's allegations as to the cause. If the report is positive, the solicitor will reapply to the Legal Aid Board, requesting an extension to the certificate to allow for the issue of proceedings and obtaining further relevant reports.

The other limitation on a legal aid certificate is a financial one. Certificates are now always issued with a financial limitation. If a certificate is limited to, for example, £7,500 of work, a solicitor should not undertake work to a value in excess of that amount without first going back to the Board for authority. This means that solicitors now have to keep a close eye on the level of their own costs, and any other fees that they have incurred (for example, experts' or barristers' fees), as the case proceeds.

When an application is made to extend the scope of the certificate, either in terms of the work that can be done under it or its financial limit, the solicitor is again required to give an assessment of the strength of the case and its prospects of success. There is, therefore, a continual reassessment of whether or not the legal aid certificate should be allowed to continue.

It is very important, therefore, for the solicitor to keep a close eye on the costs being incurred to ensure that, once a certificate has been granted, they do not spiral and make the claim uneconomic. Where a case is very strong, the limitations on what work can be done and the financial limitations on the certificate will be less stringent.

Discharge of a legal aid certificate

If, after investigating a claim, it becomes apparent that it will be extremely difficult (if not impossible) to prove the claim, or that there is little prospect of success, or the costs of bringing the claim outweigh the benefits to be gained, then it is possible for the Legal Aid Board to make a decision to discharge the legal aid certificate. If this decision is made, the legally assisted person will be given an opportunity to make representations to the Legal Aid Board as to why his/her certificate should not be discharged. These representations can be made orally or in writing. Where the solicitors disagree with the Board's assessment, they will usually help the legally assisted person with these representations. If, after all, the legal aid certificate is discharged then the case will have to be discontinued. Solicitors are entitled to recover their costs from the Legal Aid Board. Assisted persons are not liable for the costs, except to the extent that they may have already made contributions to their legal action from either income or capital.

Where a legally aided claim is successful

Legal aid should be seen as a loan from the Legal Aid Board to the assisted person that is repayable where the case has been successful but written off when the case is lost.

As was explained above, the general rule in litigation is that the loser pays the winner's legal costs. This means that in a successful claim for environmental nuisance, for example, the polluter may be ordered by the court to pay damages for past nuisance, to prevent further nuisance occurring in the future and to pay the plaintiff's legal costs. The costs will be dealt with according to procedures fixed by the court or they can be agreed between the parties. All the plaintiff's costs, which were met by the Legal Aid Board in funding the claim, must be repaid to the Board, and should come largely from the moneys paid out by the defendants. However, more often than not there is a shortfall between the costs that the losing party is ordered to pay and the actual costs that are incurred. In that case, the Legal Aid Board has first call on the damages from which to recoup any unrecovered costs. This is what is known as the 'legal aid statutory charge'.

In most cases, the charge will only be a small proportion of the damages, unless there have been problems with the case that have resulted in a high level of unrecovered costs.

An example might be a claim for nuisance where 50 people have been successful in proving their claim and have each been awarded £5,000 damages plus costs, making a total of £250,000 in damages. The total costs paid by the Legal Aid Board to fund the action are £100,000. The defendant is ordered to pay £90,000 towards the costs leaving a shortfall of £10,000. This shortfall has to be paid by the 50 plaintiffs from their damages, each paying a proportion, which in this case would be £200 each, leaving them with £4,800 each.

Where a legally aided party loses the case

Although the usual rule in litigation is that the loser pays the winner's costs, where one of the parties is in receipt of legal aid, the court will make an order that they should pay the costs but that the order should not be enforced without the court's specific permission. This is often known as the 'football pools' order because the court will usually only order a legally aided plaintiff to pay the other side's costs if they receive a sudden windfall.

Where many people are affected by pollution

In environmental pollution claims, there are usually a number of people affected by the problem. Many environmental pollution claims, particularly nuisance actions, will therefore be 'multi-party' actions. The concept of group litigation is something that has become increasingly significant in the British legal system over the last ten to fifteen years, owing to the number of claims related to mass disasters, drugs and now environmental pollution.

The British courts do not recognise the concept of 'class actions' that exists in the US. However, they have come to realise that multi-party actions need to be dealt with differently from ordinary cases where only one or only a handful of people are bringing an action. A series of procedures has been developed to deal with multi-party cases, some of which relate to their funding by legal aid.

In June 1992, the Legal Aid Board issued a special procedure for dealing with multi-party personal injury actions. The Board's definition of a multi-party action is that there must be at least ten legally aided cases, where the cases have common issues to be determined and where the claims involve a personal injury element.

Even in what are strictly regarded as nuisance actions, there will often be a claim for personal injury damage, and it is likely, therefore, that the multi-party action provisions will come into play. The procedures are designed to simplify the system for dealing with multi-party claims, to avoid bureaucracy and cost.

Once the common certificates have been issued, a special committee of the Legal Aid Board will decide whether the action meets the Board's multi-party criteria and whether the case is suitable to be dealt with under the Board's special arrangements. If the Board accepts that an action fits the criteria, any of the law firms with one of the legally aided clients will be invited to submit a tender to the Board, setting out how they would conduct and coordinate the whole action.

Until the tender process has been completed, it is possible that only a limited amount of work will be undertaken in the claim. Once the tender process has been completed, it is usually the case that one or a small number of firms will be given the job of preparing all the cases for trial, and a number of other firms may possibly be given the responsibility of organising the individual clients. The firm initially instructed to deal with the claims will not necessarily be awarded the contract to do the work.

OTHER WAYS OF FUNDING LEGAL ACTION

Legal expenses insurance

With cutbacks in eligibility for legal aid, which have left a significant number of people ineligible for legal aid, more people are taking out legal expenses insurance. This is sometimes obtained as a separate item of cover or, alternatively, might be available as part of a household or car insurance policy.

Where individuals are contemplating taking a legal action, in assessing the possible methods of funding a legal action it is useful to consider whether any of them have previously taken out legal expenses insurance when paying for their household or car insurance. As it is often the case that people forget that they have taken out legal expenses insurance as part of their household policy cover, it is important to ensure that all the individuals check their policies specifically to see whether this cover has been obtained.

If an individual is covered by legal expenses insurance, it is possible for them to contemplate funding their case privately. The legal cover available is usually in the region of £25,000 to £50,000, to cover both legal fees and the costs of experts. The fine print of the insurance policy has to be considered in each individual case.

Sometimes the insurance policy will cover a certain amount of legal costs to be incurred by the individual's solicitor, with an equivalent amount of money to be assured if the case fails and the individual is ordered to pay the other side's costs. However, some policies do not cover the defendant's costs if the claim is lost and an order for payment is made. If that is the case this greatly reduces the potential use of legal expenses insurance and the plaintiff is running the risk of having to pay the other side's costs if the action is unsuccessful.

Although the company that runs the legal expenses insurance will suggest that their policy-holder should use a solicitor who is known to them, because of the application of a European directive, Council Directive 87/344/EEC.22.6.87 Article 4, the insured can insist on using the solicitor of choice.

If a person has legal expenses insurance then the insurer acts in a similar role to the Legal Aid Board, allowing certain steps to be taken and then requiring the solicitor to report back before authorising further expenditure.

No-win no-fee agreements

These are similar to the no-win no-fee agreements, known more formally as contingency fee agreements, used in the US, but in the UK they are known as 'conditional fees' and are a hybrid of the contingency fee system. In the US, the lawyer's fee is based on a percentage of the damages that a plaintiff is awarded and is typically set at one third of the level of damages, although this might differ from case to case. Permission for cases to be undertaken in the UK on a no-win no-fee basis was introduced in June 1995 by Parliament. The system applies only to bringing personal injury actions, insolvency claims and cases heard in the European Court of Human Rights.

The no-win no-fee system allows lawyers to take on cases without charging their clients. The client and the lawyer decide whether or not the disbursements such as experts' and doctors' fees are to be funded by the client as the case progresses, or to be included in the no-win no-fee agreement. Barristers' fees can be treated as a disbursement that is funded by the client throughout; alternatively, the barrister can also take the case on a no-win no-fee basis or the solicitors can undertake to pay counsel's fees themselves.

The incentive for lawyers to take on the risk of conducting a case on a no-win no-fee basis is that they can charge their client a 'success fee' in the event that the client wins the case. This is not a percentage of damages, as with the system in the US, but an enhancement of the solicitors' and, where appropriate, the barristers' costs. The success fee will be paid out of the client's damages but is not calculated by reference to the amount of the client's damages award.

The success fee can be anything up to twice the normal fees that would be charged. The level of the success fee should reflect the degree of risk that the lawyer is taking on and the fact that the lawyer only gets paid, if at all, at the end of the case. Therefore the riskier the case, the higher the success fee, but with a cap of 100 per cent.

The Law Society's model agreement includes a 'cap' on the success fee where it is suggested that the total success fee should be limited to no more than 25 per cent of the damages recovered. This is a recommended and not a statutory limit. There are, therefore, two caps on the amount of the success fee: (1) the specific agreement as to the percentage of costs recovered, up to 100 per cent maximum, and (2) an optional but recommended additional cap of 25 per cent of damages.

Take, for example, a claim where the damages are worth £100,000 and the client and the solicitor think the chances of success are reasonable. The solicitor offers the client a conditional fee agreement with a success fee of 25 per cent plus a further 5 per cent because the solicitor is to fund the disbursements throughout the case. The case is successful and the client is awarded £100,000.

Assuming the plaintiff's costs are £25,000, the plaintiff's solicitor will largely, if not entirely, receive those from the defendants, the plaintiff's solicitor is then entitled to charge an additional success fee of 30 per cent of those costs, that is £7,500, which the client is required to pay out of his/her damages, leaving the client with £92,500.

However, take an example where the damages are only £10,000 and the costs have also reached £10,000 and where the success fee is 30 per cent. There, the success fee is £3,000; however, because of the application of the '25 per cent of damages cap', the client will only be required to pay £2,500 of his/her damages, that is 25 per cent of the damages.

The great advantage of the scheme to clients, especially where they are outside the legal aid eligibility limits or have been asked to pay a high contribution throughout the duration of their certificate, is that for a premium of £85 they can buy an insurance policy that will cover them for up to £100,000 of the other side's cost and any disbursements that they incur, such as experts' fees. The policy will only come into operation if the case is lost or has to be abandoned after the issue of court proceedings. If the case is either lost or abandoned, the solicitor and barrister who have agreed to conduct the case on a no-win no-fee agreement will not be paid.

The insurance policy will enable clients to sue where otherwise they may have been put off by the rule that the loser pays the winner's costs. The insurance policy is called 'Accident Line Protect'. It only applies to personal injury cases other than those involving medical negligence. The insurance policy does not apply to cases of personal injury that are also multi-party action cases, or to any claims involving drugs or tobacco.

In order to obtain the Accident Line Protect policy, the individual must instruct a firm that is a member of the Law Society's Accident Line scheme. The insurance company offering protection has no discretion to refuse a particular individual's insurance, but proceeds on the basis that if the lawyer agrees to do a case on a no-win no-fee basis there must be a strong probability of success.

Although conditional fees have only been up and running for a relatively short while, it is clear that, after a slow start, they will make a significant impact on personal injury cases. It is estimated that in the first year about 1,000 conditional fee agreements were entered into.

It is clear that many lawyers are reluctant to take more complicated cases, such as those that arise from exposure to toxins or pollution, on a no-win no-fee basis, for fear of failure. Some lawyers, however, are prepared to take these risks. Clearly the reduction in legal aid eligibility levels indicates that there are large sectors of society that are too well-off to be eligible for legal aid but are unable to take on the risks of funding litigation privately. Conditional fee agreements for personal injury fill a gap in the market for this group.

However, other than in straightforward cases such as those involving injuries resulting from a one-off incident, it is unlikely that conditional fees will have much impact on the conduct of litigation relating to environmental pollution, where questions of causation and liability are often complex.

It is thought that, at the outset of the scheme, few lawyers will be prepared to take such cases on a no-win no-fee basis, because they, together with the barrister, will not be prepared to take on the risks associated with this type of litigation. It may well be that if the scheme proves to be a successful method of funding ordinary personal injury cases, lawyers will become more confident in their risk-assessment abilities for more complex cases.

It is useful to compare the conditional fee system in the US with that in the UK. It may be perceived that British lawyers are reluctant to take risks with their own money, whereas in the US there is no shortage of 'speculative' or high-risk litigation. The fundamental difference, however, is that the rewards available to the lawyer in the American system are much greater because the damages awarded (usually by juries) are significantly higher. It follows that the money available as a cushion against losing cases is that much greater and, accordingly, greater risk can be taken.

The Accident Line Protect insurance policy is not available in a multi-party claim, but it is still often possible to negotiate individual insurance policies with the company. These have to be assessed on a case-by-case basis and, inevitably, the policy premiums will be higher than the policy premium for a straightforward case. The amount of the policy will depend on what the insurer perceives to be the risk involved. In cases involving straightforward, one-off incidents where the causation issues are not complex and where a large number of people have been affected, there is no reason to assume that an insurance policy cannot be obtained at a reasonable cost not greatly exceeding the £85 policy. It is not yet clear whether the costs of the insurance policy can be recovered as part of the costs of the litigation.

Privately funded actions

Where individuals have substantial means, it may be possible for them to fund a legal action themselves. In these cases, they would need to be affluent enough to fund their own legal costs and those of the other side in the event that their case was unsuccessful. It is difficult to imagine many cases today where a client would opt to fund the cost of litigation privately, and run the risk of having to pay the other side's costs if the case were unsuccessful, rather than ask his/her lawyer to enter into a no-win no-fee agreement, where such an option is available.

An example of where an individual might want to bring a case on a privately paying basis is where it is so strong that he/she does not want any of their damages to be eaten up by a success fee. This might be an option in a case where an escape of gas causes an immediate reaction and liability is not in dispute. In those circumstances, where the case is almost certain to succeed, the plaintiff might think that privately funding the action is the preferred option.

In addition, it is only possible for lawyers to undertake no-win no-fee agreements for cases that have an element of personal injury. Although it is arguable that even the most minor nuisance cases have an element of personal injury, it is not entirely clear whether the Accident Line Protect insurance policy will grant insurance for cases that are essentially claims for nuisance. In any event, as stated above, this insurance will not be available in the case of a multi-party action. However, as the insurance scheme becomes more established, the main insurers are likely to want to expand the areas they cover, in which case tailor-made policies to suit individual cases, which may also be multi-party cases, might be negotiated.

The white knight

In rare circumstances, it may be possible to find a white knight – someone who has the means to offer financial backing for a claim of public importance. This happened in the Opren drug litigation, where the plaintiff's lawyers were able to obtain the support of Godfrey Bradman, a millionaire philanthropist, to back the claim. The chances of this happening are slim, but it may be worth considering.

CONCLUSION

Funding any legal action is no fun, taking account of the steep hourly rates charged by lawyers and experts, the increasing court costs and the daunting prospect of having to pick up the defendants' legal tab if the case is lost. In virtually every environmental case the costs are likely to run to tens if not hundreds of thousands of pounds. As a result, very few people will be able to take this step without some sort of support. That support is most likely to be achieved through the legal aid system, for those eligible, or through insurance.

The legal aid system is both complex and increasingly tight. It is easy for the applicant or the inexperienced lawyer to give up when trying to obtain funding in an environmental case, where the hurdles erected are undoubtedly high. The best advice that can be given to people in this position is to seek out lawyers with experience in this field.

4   The Media

Environmental actions attract a great deal of media attention. What environmental campaigners have to ensure is that this attention operates to their benefit rather than to the benefit of those being targeted by legal action, the defendants.

Almost without exception, the media start off on the side of the complainants or plaintiffs. For many journalists the David-against-Goliath fight is what brought them into their profession. However, it is crucial that the plaintiffs and their lawyers employ tactics that will retain not only the journalists' interest but also their broad support. The support of the media is one of the main weapons in the hands of the environmental campaigner. Understanding how the media work and using that knowledge is central to a successful campaign.

It is important, therefore, that the media are handled with great care. While it offers no formula for dealing with them each step of the way, this chapter describes how the media work and considers how campaigners might best use them, while remaining aware of the potential problems as well as the benefits.

Companies today seem to be taking an increasingly aggressive approach to comments they perceive to be defamatory. It is particularly important, therefore, that campaigners understand the parameters of the laws on defamation, and this chapter provides a guide to defamation to assist campaigners in this area.

How the media operate

The media are constantly evolving, and never more rapidly than now, with current affairs programmes appearing and disappearing with great rapidity and new satellite, cable and digital terrestrial stations coming on stream. With the ever more competitive nature of today's society there is also a much greater flow of journalists from one job to another. This means that contacts can alter regularly, telephone and fax numbers can quickly become obsolete, and journalists can come and go in the blinking of an eye.

It is up to campaigners to keep their fingers on the pulse of these changes. There is little point in putting time and effort into sending out a faxed press release that no one covers because it has gone to offices the journalist has long since left, or because the environmental correspondent is on holiday or has become the gardening correspondent.

In any environmental case an early decision has to be made on whether the issue and subsequent case are likely to attract national or only local interest. That is very difficult to judge. Clearly, in major disaster cases, such as occurred when the *Sea Empress* went aground off the Welsh coast, the incident and subsequent environmental actions are likely to gain national interest, at least in the initial stages. In other instances it may be very hard to enthuse journalists, particularly where the story is well away from the Southeast. Hard-nosed hacks can turn down perfectly good stories simply because they involve too much effort.

If the action only involves a handful of people who have had very minor injuries following a leak of some pollutant, the answer may be to concentrate on local publicity, particularly because the class of potential claimants is likely to be local. Where, however, the impact is much wider and people all over the country are affected, the national media are likely to be far more interested.

It is impossible to be prescriptive about how to work with the media, because there are no guarantees about how they will react to any story. What may be a brilliant story in normal circumstances may obtain no coverage at all if it is announced on the day of a royal wedding, and a story that seems to the lawyers to be rather limited may receive remarkably good publicity, perhaps because it is announced on a Sunday, or during August when stories are very hard for the media to come by.

The general tips set out below must be considered in this context. There are no guarantees that the effort put into publicising a case will result in media coverage.

Agencies

Where there is a story of national interest, which is highly likely to gain wide media coverage, or where the campaigners want to say something on a case where there is already substantial interest (a disaster would be the best example), the Press Association (PA) is the best agency to contact first. This is the news agency that most of the British media look to for obtaining hot press stories. Other agencies, such as Reuters and Associated Press, are more useful for sending stories out to the international media (see list of contacts on page 413). Whether the press release is sent out more widely depends on the time available. The national media will almost certainly pick up a 'hot' story from the PA, so sending out the release to other newspapers is probably a waste of effort.

Press release

The press release itself should be short and punchy, but give sufficient detail and include a couple of quotes from relevant people involved in the campaign. Sound bites are what the press are after, not long, turgid ecobabble. Journalists should be able to understand the story clearly from the press release, so that when they contact the campaigner it is only to check up on a few additional points. They may simply phone up to obtain a few original quotes, to ensure that the story in the *Guardian*, for example, is not exactly the same as that in *The Independent*.

There is nothing worse than being telephoned by a host of journalists on the same day and finding either that none of them has the release or that they haven't bothered to read it; repeating the same details time after time can be extremely tedious. In these circumstances it is probably best to ask the journalists whether they have the press release to hand, and, if not, to fax it through to them immediately so that they can read and digest the information before phoning again.

Which journalists?
When sending out a press release more widely than to the
agencies, it is important to make a positive decision about which
journalists, or at least which types of journalist, are likely to cover
the story. The choice is either to the newsroom of the newspaper;
to the planning desk of the TV or radio station, where the events
to be reported are in the future, or to their newsroom, where
events are occurring that day; or to a particular journalist.

The benefit of sending the release to the newsroom is that it
avoids the risk of the specific journalist being away – either on
holiday or on another story. However, the campaigner needs to be
aware that newsrooms receive thousands of press releases every
day, and that they will follow up only a tiny fraction of these.
Further, the journalist who is likely to be assigned to the story
may well be a general reporter who will start off with next to no
understanding of the issues, the law or the way these cases are
progressed.

Each general journalist who phones to follow up the story is
likely to need a good deal of time if the story is to be used and to
be correct when printed. It is important, therefore, to cultivate
contacts in the media. A journalist who knows the authors of a
press release and has respect for their judgment is more likely to
follow up a story than the journalist who receives a press release
from an unknown source. Further, journalists with a specific
remit (such as legal, health, environment, etc.) will usually be
clear about whether the story will be of interest to their editors, so
time spent with those journalists is unlikely to be wasted.

Getting on
It is still true, however, that no story is certain to be used until it
actually appears in print, in a bulletin or whatever, regardless of what
the journalist says. One of the authors of this book had the experience of being allowed into ITN to watch the *One O'Clock News* going
out from behind the scenes. At ten past one a new item was brought
into the production room. The producer saw that the piece was one
minute twenty seconds long, checked down the remaining schedule
and saw that item eight (which was a piece about Nelson Mandela),

out of the twelve news pieces due to be shown, was of the same length. That piece was therefore extracted, despite having a higher priority than a number of other news items.

Nelson Mandela is unlikely to have been heartbroken by this turn of events (and probably wasn't even aware of it), but if the same thing happened to a campaign where people had spent many hours with a film crew giving interviews, and had told all their friends and family to watch, it could have quite a dispiriting effect on the campaign.

Local media

With the local media, life is rather more relaxed. For any environmental legal action that is based locally the media are likely to be very interested, and they will generally do far more of the running on the story. They are also more likely to take stories straight off the press release, so it is important that they are kept on the mailing list. Getting to know the local reporters is not so difficult as they are not inundated with news in the same way as the national media and are more likely to give major coverage to environmental issues.

The early days

There are many reasons why it is important to be in touch with the media in the early days of a campaign. The media can help to attract supporters to a campaign. They can put pressure on defendants to reach an early settlement rather than have continued adverse publicity. They can help to convince the Legal Aid Board, and its constituent parts, of the case's public significance, which may encourage them to look positively on the action (although the reverse can also be the case). Finally, they can help to ensure that, by the time the case is tried, the judiciary is aware of the issues and have overcome any scepticism they may have had when hearing the proposition for the first time.

Advertising

One of the issues that can arise in the context of using the media in a campaign is advertising by the lawyers. This might be to build up the number of plaintiffs in the existing case to ensure the

Legal Aid Board's cost–benefit rule is met (see Chapter 3, pages 45–6), or to bring in additional cases and therefore build up the funds available for the action. Although advertising is primarily an issue for the plaintiffs' lawyers, it may well affect the way in which the case is run, so an understanding of the lawyers' concerns is likely to assist the plaintiffs' comprehension of the lawyers' actions.

The issue of advertising is a difficult one for the plaintiffs' lawyers and one where there is no unanimity of view among those experienced in this field. To advertise, whether generally or specifically, is certainly within the Law Society's rules, and has been since 1984. The Solicitors' Practice Rules 1990 simply prohibit advertising in any manner that could be seen as being in bad taste, that is inaccurate or misleading, or that falsely claims expertise.

Regardless of the position adopted by the Law Society, there is an important question here about the extent to which, by advertising, the lawyer is unreasonably building up the claimants' hopes that there is a winnable case, when the chances of success may in reality be less than 50/50. It is not the role of this book to take a particular moral stance on this issue, but for those whose lawyers do end up advertising it is important that the significance of the action is understood. Michael Napier, President of the Association of Personal Injury Lawyers in 1994, made a similar point:

> Great care has to be taken not to raise in the minds of the public expectations that cannot be fulfilled. I anticipate that there will be different views about this sensitive and controversial area of the way that solicitors behave and although we do not advertise, I do not cling too tightly to the moral high-ground because I am aware of one successful publicity campaign in this country involving compensation for many badly injured women who came forward as a result of advertisements.

There are two prime reasons for advertising. The first is to let potential claimants know that the possibility of making a claim exists. This position is most likely to arise in the areas of product liability and environmental claims. Where the lawyers have been able to determine in a particular area that because of scientific or legal developments there is a realistic prospect of pursuing successful claims, advertising can be one of the only ways of contacting those people who are victims.

In the Sellafield claim, for example, the lawyers advertised in the *Whitehaven News* (July 1988), the circulation of which covers the West Cumbrian region, including the nuclear plant and those living in the immediate vicinity. In the action against tobacco companies taken by people who alleged that they had suffered illness from smoking cigarettes, adverts were placed in Liverpool and London (for example, in the *Liverpool Echo* and the *Hackney Gazette* in June 1992) to ensure that the two firms of lawyers involved had a wide base of clients from which to choose the lead cases that would go forward.

The second reason for advertising is that the lawyers may well be keen to gain a market share of the clients in a multi-party claim that already exists – for example, following a disaster of some kind.

What is important for the campaigner to understand is that, if their lawyers advertise, the burden of responsibility placed upon their shoulders is far greater than normal. When clients walk in through a lawyer's door looking for help, it is the lawyer's role to give that help to the best of his/her ability. However, when a lawyer advertises for cases, he/she is inviting the victims, who might not otherwise look for legal help, to come forward, and is telling them that they have the makings of a case. The lawyer–client relationship, and the expectations and responsibilities involved, are therefore significantly different.

Indeed, in the Bar Council's response to the Legal Aid Board's position statement on multi-party actions, it expressed the concerns of barristers regarding the impact of advertising and the general publicity that surrounds a multi-party action, saying: 'that many may have the false impression that by joining in, they will recover large sums by way of damages' (paragraph 6).

In its response to the Legal Aid Board's paper, McKenna & Co. also considered the issue of advertising and publicity and said:

> It often emphasises the potential for compensation and that there will be no downside for the claimant because the firm will assess the availability of legal aid at a (free) first consultation organised

through a (free) phone line. This form of advertising inevitably
also increases the potential for poor cases being brought forward
as well as the good ones. The complexities of such litigation are
not highlighted. (Response to the Legal Aid Board's paper on
group actions, May 1994, page 22)

This argument would be stronger if advertising in such a way were
carried out as a matter of course in a whole series of cases. The
reality is that there have been relatively few instances of advertising to encourage victims related to specific products or incidents
to come forward (although this is a trend that is on the increase).
However, the general message put forward by McKenna & Co.,
that there is a downside to advertising, should not be ignored.

Campaigners should ensure, therefore, that if their lawyers
advertise in this specific way they must have carried out sufficient
investigations to be clear that the claim genuinely has a solid base
and that the potential clients have at least a reasonable chance of
success. Indeed, the British Code of Advertising Practice demands
that adverts should not be misleading or contain exaggerations,
although it does allow for the free expression of opinion (Annex
11C, clause 5.2).

Sub Judice and Contempt

The Contempt of Court Act 1981 created strict liability if a
publication creates a substantial risk that the proceedings will be
seriously impeded or prejudiced.

The case has to be 'active' or *sub judice* for the restrictions to
apply. The *sub judice* period is the period when the matter is
primarily before the courts, and they frown upon the issues in the
case being considered in public during this time. In civil cases this
means that the case has been set down for trial or a trial date is
fixed. However, the case will also have prior 'active' periods
relating to interlocutory (i.e. procedural) applications lasting from
the time when the hearing is fixed to the date it is heard. The
proceedings end when the matter is resolved at court or in a
settlement or is discontinued. Appeal proceedings are active from

the time the notice of appeal is lodged until after the appeal is disposed of.

Civil cases on the whole receive less attention than criminal cases and because juries do not hear civil trials there is a minimal risk that media coverage will influence the outcome. However, environmental actions generate intense media interest and campaigners need to be aware of the rules.

The risk of contempt may arise in numerous situations. During the Lockerbie Fatal Accident Inquiry the *Sunday Telegraph* published an article that criticised the plaintiffs' steering committee and counsel for their narrow line of questioning, which was said to be motivated by the compensation claims. Sir Max Hastings and the journalist concerned were quickly brought before the court, where it was held that the article was not in contempt of court. Contempt proceedings can be initiated by the judge or the court that is affected but more usually the matter is referred to the Attorney General, the chief legal officer in England.

In the 1960s a group action arose as a result of the prescription of the drug thalidomide as a tranquiliser to pregnant women. Many of the children born to those women had very severe birth defects (*S and Another* v. *Distillers Co. (Biochemicals) Ltd* [1969]). The *Sunday Times* ran a campaign alongside the court actions, every week castigating the defendants, Distillers, for fighting and not settling the claims. The court was critical of this campaign, indicating that Distillers had the right to defend themselves, but in the end the action settled.

A more recent case where this has been an issue is the Greenpeace action against the government, which is reported in the law reports as *R.* v. *Secretary of State for the Environment* et al. ex parte *Greenpeace Limited and Lancashire County Council* (1993), when they were attempting to overturn the decision of the Secretary of State to allow the commencement of the new Thorp nuclear reprocessing plant in a judicial review application. British Nuclear Fuels was a party to the proceedings as a party directly affected by the outcome. Its in-house lawyer, Alvin Shuttleworth, was taken to task in court on the second day of the hearing by the judge, Mr Justice Potts, who had seen him on television the night before criticising Greenpeace for having brought the action in the

first place. The ticking-off, however, was the only action the judge
was prepared to take.

While campaigners and their lawyers should always be careful,
the courts are unlikely to be concerned about media appearances
that are simply explaining the basis of the action. Fair and
accurate reporting is a defence to contempt of court. What should
be avoided is heavy criticism of the defendants, or attempts to be
too contentious. The reporting of interlocutory applications held
in chambers should be dealt with carefully. The outcome of such
applications, if of general interest, should be revealed in open
court but, if it is not, clearance from the judge should be sought.

Further it is probably unwise for campaigners and their lawyers
to give interviews once the trial has commenced. The judge may
well become unhappy if they are seen to be giving a running
commentary on what is occurring in the courtroom. This often
happens in the US, and is increasingly becoming an issue in
criminal cases in this country.

The prospect of being held in contempt of court because of
statements made to the media is therefore more of a concern than
the use of the *sub judice* rule, partly because to be brought before
the courts in such an action would be far more damaging than a
sub judice complaint, and partly because the courts take contempt
far more seriously.

For campaigners and lawyers involved in this type of action there
is a continuous pressure from the media to reveal new facts arising
from the case. Where the defendants provide evidence of a startling
and newsworthy nature through the discovery process, where documents are produced that are relevant to the case, there might be the
temptation to broadcast this information. The temptation should be
resisted. Indeed, in many cases the defendants obtain undertakings
from the plaintiffs' lawyers specifically agreeing to the restriction of
the use of the more contentious documents.

The basic rules of contempt are that documents arising from
the defendants (other than experts and counsel who are working
on the case) cannot be revealed to a third party without the
specific authority of the defendants. This restriction applies to the
passing on not only of the document itself but also of information
arising out of the document.

The same restrictions apply to documents gained from third parties by way of the discovery proceedings.

During the course of the trial, the principle is that once a document such as an affidavit (a sworn statement) has been read out it can be passed on to the media. This does not, however, apply to any documents, discovered from the defendants, to which the affidavit refers. In the Harriet Harman and Home Office case, Ms Harman was held to be in contempt for passing over to a journalist a document from the Home Office, despite the fact that the contents of the document from the defendants had been read out in open court. Although the law was relaxed after her case to allow disclosure in these circumstances, the general strict rules regarding disclosure still apply.

The position regarding the disclosure of pleadings to the media is more equivocal. The writ is a public document, access to which can be obtained by any journalist from the High Court's writ room, and therefore passing on that document to the media is in no way a contempt. Several multi-party actions where protective writs were issued but not served were featured in the press after writ searches had been carried out by commercial companies (for example, the British Airways flight BA 149 hostage situation at Kuwait airport and the PIA air crash in Nepal).

The rest of the pleadings (that is, the formal legal documents setting out the case of the two sides) are, however, another matter. The question is whether revealing any of the pleadings will reveal discovered documents. If, therefore, one of the pleadings makes heavy reference to a discovered document, there is a greater prospect that the passing of this on to the media could be held to be in contempt. Where this is a possibility it is important for the campaigner to take specific advice from the solicitors and, possibly, counsel, bearing in mind that the lawyers will usually take an extremely cautious line.

In general terms, the campaigners and their lawyers should see the media as an asset to be worked with for the benefit of the client. In saying this, the perspective of the defendants is clearly very different. In their paper responding to the review by the Legal Aid Board of multi-party actions referred to above, McKenna & Co. made the following statement:

The effect of media coverage of health issues cannot be underestimated. Whilst potential claimants tend to lack knowledge in the field relevant to their claim and in the past such ignorance may have discouraged the bringing of claims that might have been reasonable, circumstances today are very different. The media's interest in health tends to emphasise safety problems on the one hand or alternatively raises unrealistic expectations of benefit with the latest drugs and research. Media publicity tends to be sensational and does not educate on the fundamental issue of risk/benefit, particularly with the new products developed at the frontiers of science. Such publicity encourages claims and then publicity concerning the claims themselves leads to more claims creating a classic 'bandwagon' effect, where more and more people who have taken a particular product are encouraged to believe that they may have a claim, and feel reassured about the merits of the apparent numbers of people similarly effected. (Response to the Legal Aid Board's paper on group actions, May 1994, page 22)

While plaintiffs' lawyers are unlikely to agree with much of this statement, it is important to understand where the defendants are coming from in their perception of the media's role.

Defamation

Many campaigning organisations have become increasingly cautious about the statements they make alleging environmentally unsound practices by companies, since in recent years some companies have adopted policies of hinting at or threatening defamation actions in order to stifle and intimidate critics and campaigners, particularly in areas where the facts are controversial or not widely known.

Defamation can be defined as 'the publication of material that reflects on a person's reputation so as to lower him/her in the estimation of right-thinking members of society generally'. Thus to be defamatory the material published must in some way injure the plaintiff's reputation, and a false statement against a company will not necessarily be defamatory if it does not do so. A corporation may, however, have a different remedy, called 'malicious falsehood', without a lowering of reputation. It would need to prove that a false statement was made spitefully, dishonestly or recklessly and there would usually need to be financial loss resulting from the statement. A corporation may only sue for defamation in respect of statements that damage its business reputation (criticisms made on environmental grounds are likely to fall into this category). Local government bodies cannot sue in defamation at all, although individual officers or employees singled out for criticism will have a potential right of action.

The defamation action will either be called 'libel', if the statement is written down, in films or in videos in permanent form, or 'slander' if it is verbal. Plaintiffs may sue for libel even though there has been no financial loss, but for slander there must generally have been actual damage suffered.

Any party has a right to trial by jury, unless the court decides against this on the basis that 'prolonged examination of documents or accounts or any scientific or local investigation which cannot conveniently be made by a jury' is necessary. Plaintiffs must normally commence their action by issuing a writ within one year of the publication of the statement in question. Defamation is the only important civil right for which legal aid is not available. This means that wealthy plaintiffs can potentially inflict crippling legal costs on campaigners, and those operating with scarce resources are likely to be denied legal representation.

Judges or juries place themselves in the position of 'the ordinary reasonable reader' of the publication in question, and decide on this basis whether they think that the statement made would injure the plaintiff's reputation. There are a number of legal rules used to interpret the potential meaning and effect of a statement. For example, the courts do not just consider a document on its face value but will consider whether a person reading the statement would be able to 'read between the lines' and understand the real meaning even if that meaning is not spelt out.

The plaintiff initially has the burden of proving that the statement made is defamatory. Once this has been done, however, the defendant then has the burden of proving that he/she has a valid defence: for example, that the statement was true, or was an honest comment made on the basis of true facts. Thus the statement is presumed by the court to be false, and though it may be true the defendant has to go to the expense of proving this in court to prevent damages being awarded against him/her. Furthermore, the defendant cannot use 'hearsay' evidence such as books and press reports to do this, but must rely on official documents and direct witnesses. This may be extremely difficult for the defendant if, for example, the source of the information is unavailable, or a witness has died or wishes to remains anonymous.

If the defendant can show the truth of the statement, then this acts as a complete defence, known legally as 'justification', to any defamatory statement of fact, whatever the motives for its publication. It is not necessary to prove that every single fact is true, so long as the 'sting' – that is, the most defamatory part of the statement – is true. There are, however, restrictions on how far from the truth the statement can stray. For example, a false, generalised statement that is made on the basis of a true single incident will not be legally justified. Where a defamatory statement is published, liability cannot be avoided because the statement is expressed as having been made by someone else, for example 'We have been told that ...'. An important advantage for the defendant is that he/she will be able to cross-examine the plaintiff in court in detail about all the matters that are relevant to the truth of the statements made.

The other main defence to defamation is that of 'fair comment'. This arises where the defendant makes an honest expression of opinion on the facts that are known to him/her. No matter how unfair the opinion, he/she cannot be challenged if it is honestly made. The question that the courts will ask is whether the views could honestly have been held by a fair-minded person on facts known at the time. Whether or not the jury agree with the statement is irrelevant. This defence only relates to 'comment', in other words statements of opinion, and not to statements of fact. A defamatory statement of fact must be 'justified', that is, proved true, which is generally a more difficult task. In practice it is often unclear on the face of it whether a statement made will be considered fact or opinion.

The test used by the courts is an objective one of what the 'ordinary reader' would consider the statement to be.

Unattributed statements published by campaigners are likely to be considered factual, and preceding words such as 'In our opinion ...' will not necessarily be enough to rebut this. The defence of fair comment will not succeed if the comment is made without a factual basis; it must be based on something that is true or substantially true.

Another defence of 'qualified privilege' may be available to campaigners in very limited circumstances. Any statements that have been made from what the court considers a social or moral duty cannot be subjected to an action for defamation, on the condition that they are made honestly. However, the court does not consider that a 'social or moral duty' will arise in a wide range of circumstances. For example, it has been reluctant to hold that there is a 'moral' duty on journalists or campaigners to publish 'fair information on a matter of public interest'. In circumstances where there is an alternative method of seeking redress, the court is unlikely to find such a duty, and it is only in extreme situations such as suspected terrorism that this defence could be used. As environmental awareness grows, however, it may be possible to argue that severe environmental harm could constitute such a situation.

Damages for defamation are generally determined by juries and they are notorious for their size and unpredictability. Awards amounting to hundreds of thousands of pounds are commonplace, often bearing little relation to damages awards for, say, physical harm. Recently, the Court of Appeal has given guidance to courts on reasonable levels of damages, with reference to cases of physical harm, and this has led to some reduction in the level of damages awarded. A refusal to correct or apologise for an obvious mistake will increase the damages awarded, as will the seriousness of the defamation and the degree to which it is repeated. On the other hand, prompt apologies can lower the level of damages. Damages are intended generally to compensate the plaintiff for damage to reputation, injured feelings and for any loss of earnings resulting from the defamatory statement.

Libel cases often have the result for campaigners of creating publicity for their cause. The most famous case of this is the mammoth battle between the $30 billion a year McDonald's Corporation and two supporters of a small campaigning group called London Greenpeace, with a combined income of £7,000.

McDonald's has a history of using libel laws aggressively against its critics, and a number of organisations have had to apologise or back down, including the BBC, the *Guardian, The Independent* and Prince Philip. In this case, McDonald's sued five campaigners for libel over a six-page factsheet produced by London Greenpeace entitled 'What's Wrong With McDonald's?'. Three of the campaigners apologised, but the remaining two refused to back down, and acted as litigants in person, organising their own defence. They received limited free legal support and organised their own witnesses and documents.

The trial began in June 1994 and by November 1996 had become the longest trial of any kind in British history. It received worldwide media coverage and around 180 witnesses gave evidence on the effects of the company's operations on human health, the environment, farmed animals, the Third World and McDonald's own staff.

In June 1997 the court delivered its 1,000-page judgment, dealing with each of the allegations in the factsheet. The judgment stated that some of the criticisms made of McDonald's were true, such as the allegations that McDonald's exploits children in its advertising and was responsible for cruel practices when rearing and slaughtering animals.

McDonald's won £60,000 damages against the defendants, but have agreed not to pursue a claim against them for it. The defendants have considered challenging the right of a multinational to sue against comments made in the public interest, and also a challenge in the European Court of Human Rights of their denial of a trial by jury.

In many cases the prospects of large, unpredictable awards of damages combined with the burden of proving the truth of published statements acts as a severe restriction on the freedom of campaign groups to expose destructive and immoral practices, particularly those with substantial funds that they prefer not to be spent on lawyers' fees. However, as the McDonald's case demonstrated, a libel case can lead to worldwide publicity for a cause in a way that could never have been possible otherwise. This case will undoubtedly mean that high-profile corporations will think carefully before commencing libel proceedings and risking international exposure and cross-examination on issues that they would prefer not to be published.

CONCLUSION

The environment remains an area of great interest to the media and concern to the politicians. Interesting them in the plight of local people and of the wider issues involved should be seen as fundamental to any campaign. An issue that is splashed across the pages of every national paper will gain the interest not only of chief executives but also of shareholders and fund managers. This is where companies have their Achilles' heel. Experienced environmental campaigners recognise the media as their lifeblood. Local groups should take their cue from them.

5 Environmental Information

Nowhere is the old maxim 'information is power' more appropriate than in relation to detailed facts on the state of the environment.

Much of the information on the state of the environment is of a very scientific nature. As such, it can be difficult, expensive and time-consuming to collect. Consequently, only on relatively rare occasions are members of the public in a position to gather their own information.

Fortunately, in recent years there has been a great increase in the amount of information collected and recorded on the state of the environment and the effect on it of particular activities. Much of this information is compiled by the relevant regulatory authorities (such as the Environment Agency or the local authority) or central government.

The public has the right to some, but not all, of this information. While it is often not provided in a particularly user-friendly form and getting hold of it can, in practice, involve negotiating a bureaucratic maze, this information can nonetheless represent an important resource to an effective campaigner. The information can be used in several ways.

- *To demonstrate that a problem exists.* From global issues such as damage to the ozone layer to very local problems such as pollution of a particular stream, it can be extremely difficult to convince people that a problem exists unless the person has the information to back up his or her argument.

- *To allow companies or technologies to be compared.* Information on the environmental performance of different companies can show one to be performing in a more 'environmentally friendly' way than another. In addition, information on different technologies can show one to be preferable to another.

- *To ensure that the regulatory authority is doing its job.* Information on pollution from a particular industrial site can show whether a company is operating legally (within its authorisation, licence, etc.). If it is not, a person can then ask the relevant regulatory authority why it is not prosecuting the company. Often the threat of making a private prosecution has a galvanising effect on the regulatory authority and prevents it from taking an unduly conciliatory line with industry.

- *To enable individuals to take enforcement action themselves.* If the regulatory authorities fail to take action to stop a company operating illegally, individuals can step in and prosecute the company themselves. Central to any criminal prosecution will be information showing whether the company was operating outside its authorisation, licence, etc.

- *To provide evidence for a civil claim.* Even if a company has not breached its authorisation, licence, etc., its actions may still be causing damage or nuisance. If so, individuals can bring civil claims against the company. Again, information on the pollution produced by the company is likely to be central to any such civil claim.

HISTORY

Historically, the public has had very little access to environmental information. In most cases, regulatory authorities needed the permission of the companies that they were regulating before they were able to release any information about the environmental performance of the companies. Obviously, companies were not in a great hurry to provide the public with any incriminating material, and so very little information was released.

Even in the small number of cases where regulatory authorities were allowed to release information without companies' permission, the procedure was so complex and cumbersome that very few authorities bothered to set up the necessary administrative systems. In addition, until comparatively recently the Official Secrets Act prevented central government from releasing much of the environmental information it held.

Throughout the 1970s and 1980s, the Royal Commission on Environmental Pollution (RCEP) produced a series of damning reports criticising the culture of secrecy and recommending that the public should have far greater access to environmental information. However, after intense lobbying by industry, the government failed to act on the RCEP's recommendations. The Confederation of British Industry (CBI) was particularly vocal, with its doom-laden prophecies of how allowing greater access to information would have a catastrophic effect on the success and profit of its members. It even suggested that very modest proposals for access to certain information on water pollution contained in the Control of Pollution Act 1974 would lead to the wholesale disappearance of the chemical industry from the UK.

The arguments against disclosure were: that incorrect interpretation would be put on any information by an army of green fanatics; that there would be an increase in the legal proceedings for pollution offences; that commercially sensitive information would be released; and that the administrative costs involved in satisfying these requests would be excessive (and would eventually have to be paid for by industry itself).

It was not until the late 1980s and early 1990s that the government began to move towards a policy of allowing greater access to environmental information. This was achieved in two main ways. First, there was a wealth of new legislation relating to environment issues such as water pollution, waste disposal and pollution from industry. This was partly in response to greater public awareness of, and concern for, the environment and partly to implement requirements imposed by European legislation. Provisions for the release of environmental information were included in many of these statutes.

Second, the European Council Directive on the Freedom of

Access to Information on the Environment was adopted in 1990 and required member states (including the UK) to incorporate the requirements of the Directive into national law by the end of 1992. The Directive was implemented in the UK by the Environmental Information Regulations 1992.

The Directive and Regulations provided for the release of environmental information, generally, and therefore complemented and extended the provisions for release contained in the specific statutes. We shall now examine these two bases for greater public access to information.

STATUTORY PROVISIONS FOR ACCESS TO INFORMATION

Many of the provisions relating to public access to information are contained in the statutes dealing with the particular environmental issue. For example, the Water Industries Act 1991 and the Water Resources Act 1991 are the main pieces of legislation dealing with the regulation of water pollution and each of these contains sections dealing with public access to information.

Water pollution

Water pollution registers
Section 190 of the Water Resources Act 1991 requires the Environment Agency (or formerly the National Rivers Authority) to maintain public registers containing details of water pollution.

The information available on these registers includes details of consents to discharge pollution, applications for discharge consents, enforcement action taken and any convictions for water pollution offences. The registers will also include details of pollution monitoring. Those details will usually contain information about who took the samples, the date and time of the sampling, the maximum level of the pollutants allowed and the actual level found, thereby indicating whether a particular consent or authorisation has been exceeded.

Further details of how these registers work is contained in the Control of Pollution (Registers) Regulations 1989. These registers have to be open at reasonable times and the public can inspect the register without charge and take copies at a reasonable charge. The register must also contain details of applications for discharge of pollution into waters and sea within 28 days of the application being made and must similarly provide details of any consent given by the regulatory authority.

Trade effluent registers

Section 196 of the Water Industries Act 1991 requires the sewage undertakers (usually the privatised water companies) to maintain registers of consents relating to discharges of trade effluents to the sewerage system.

A similar provision in Section 195 of the Water Industries Act 1991 requires the Secretary of State for the Environment (now Environment, Transport and the Regions) to hold the trade effluent consents made by the private water companies.

Abstraction licence registers

Section 189 of the Water Resources Act 1991 requires the Environment Agency to maintain a register on abstraction licences as well as details of any impounding of licences.

Waste pollution

Register of waste-management licences

Section 64 of the Environmental Protection Act 1990 (EPA 1990) requires the Environment Agency to maintain registers containing a wide range of details relating to waste management. The Waste Management Licensing Regulations 1994 provide details on the information to be kept on these registers.

This information includes details of waste-management licences, applications for waste-management licences, revocations or suspensions of licences, certificates of completion granted, convictions for waste-management offences, enforcement and remedial action taken and details of pollution monitoring.

However, sections 65 and 66 of the EPA 1990 allow for some information to be excluded from these registers. Excluded information includes that relating to prospective or actual legal proceedings, on grounds of national security, and information that is commercially confidential. However, the onus is on the person applying to have information withheld from one register to convince the Environment Agency that disclosure of the information would unreasonably prejudice their commercial interests. If the applicant is unsuccessful then he/she can appeal to the Secretary of State for the Environment, Transport and the Regions.

Information that has been classified as commercially confidential is excluded from the registers for four years, although the person can reapply after this time for the information to continue to be thought of as commercially confidential.

Registration of waste carriers
The Control of Pollution (Amendment) Act 1989, and the Controlled Waste (Registration of Carriers and Seizure of Vehicles) Regulations 1991 made under the Act, require the Environment Agency to maintain public registers of carriers authorised to carry waste.

Industrial pollution

IPC/LAAPC registers
Under Part 1 of the Environmental Protection Act 1990, the most polluting industrial processes are subject to 'integrated pollution control' (IPC) and regulated by the Environment Agency, while air pollution from less polluting industrial processes is subject to 'local authority air pollution control' (LAAPC) and regulated by local authorities.

Section 20 of the EPA 1990 requires that information on these processes should be maintained by the Environment Agency or local authority on public registers. Details of the operation of these registers are contained in the Environmental Protection (Applications, Appeals and Registers) Regulations 1991.

The Environment Agency is required to maintain registers

containing information on IPC processes, while local authorities are required to maintain registers containing information on LAAPC processes and also IPC processes (from information provided by the Environment Agency) in their area.

The registers include information on applications, authorisations, variations, revocations, appeals, convictions, monitoring data, etc.

There are provisions for the exclusion of information on the grounds of national security or commercial confidentiality. The onus is on the applicant applying to the Environment Agency of local authority for the exclusion of the information on the grounds of commercial confidentiality, although an unsuccessful applicant can appeal to the Secretary of State for the Environment, Transport and the Regions. There is a power for the Secretary of State to require that the information be included in the register even if it is commercially confidential. Commercial confidentiality exclusions only last for four years, although a further application can then be made.

Chemical Release Inventory (CRI)

In addition to the IPC registers, the Environment Agency also publishes the Chemical Release Inventory (CRI), an annual report and database on the aggregate release of pollutants from all IPC processes. This information is gathered from the monitoring data kept on the IPC registers and covers nearly 500 different pollutants.

Unfortunately, no information is given on the release of pollutants by individual companies. This contrasts with the comparable database in the US, the Toxic Releases Inventory (TRI), in which data on individual companies are given in the form of 'league tables'. In the US, the emission of pollutants has been reduced by almost 50 per cent since the introduction of TRI in 1987 and it is widely acknowledged that the tables have provided industry with a major incentive to reduce emissions.

In addition, there is a huge variation in the information collected on pollutants at different sites. For example, one incinerator may be required to monitor over 40 different pollutants while another may only have to monitor five. Unfortunately, this does not necessarily mean that the second incinerator is not producing these other pollutants but merely that they are not required to be monitored. This means that they are similarly not included in the CRI database.

The first annual CRI report was published in 1994 and included information gathered in 1992 and 1993. There are plans to extend the scope of the Inventory to include LAAPC processes, discharge consents and non-point sources (such as pollutants from motor cars).

Friends of the Earth have put the CRI on their Web site at http://www.foe.co.uk/cri.

Air pollution

As discussed above, air pollution from industry is regulated by the Environment Agency or local authorities under Part 1 of the Environmental Protection Act 1990. The EPA 1990 requires these bodies to maintain public registers of information.

In addition, section 34 of the Clean Air Act 1993 gives the local authorities certain powers regarding the release of monitoring information relating to air pollution in their area, although that Act makes it clear that this should exclude information considered to be a trade secret.

Planning

Section 69 of the Town and Country Planning Act 1990 requires the planning authority to keep a public register of all planning applications and decisions. In metropolitan areas, the planning authority is the Metropolitan District Council, while in non-metropolitan areas, the planning functions are split between the County Council and the District Council.

Section 188 of the Town and Country Planning Act 1990 requires the planning authority to maintain a register detailing all enforcement action that they have taken. This would include all enforcement notices, stop notices and breach of condition notices issued by the authority (see Chapter 7 for further information on planning).

Contaminated land

Section 78R of the EPA 1990 (introduced by the Environment Act 1995) will require the relevant enforcing authorities (either the Environment Agency or local authorities, depending on the type and seriousness of the contamination) to maintain registers containing information on contaminated land.

Government policy on identifying and cleaning up contaminated land is still in its infancy and the regulations specifying the form of these registers and the type of information to be kept on them have not yet been issued. However, sections 78S and 78T of the EPA 1990 explicitly require information relating to national security and commercial confidentiality to be excluded. Observers do not expect these registers to be working for the next couple of years.

Local government

Local authority decisions are, at least in theory, made by elected local councillors, usually through the appropriate committee (for example, the Planning Committee or the Environment Committee). However, in reality most of the decisions are made by officers of the local authority, with the councillors often doing little more than rubber-stamping decisions at the relevant meetings. Usually, the officer will prepare a report for the committee on a particular issue to be discussed at a meeting, setting out the background, the various options and a recommendation.

The Local Government (Access to Information) Act 1985 allows the public to attend these meetings and also to have access to agendas, minutes and reports (including any 'background' papers cited in these documents). These provisions allow for the release of much information held by local authorities relating to environmental issues, including planning and mineral matters, environmental health matters, waste-disposal issues and air pollution issues.

However, there are grounds for excluding the public from these meetings and withholding certain information (for example, if publicity would be prejudicial to the public interest).

ENVIRONMENTAL INFORMATION REGULATIONS 1992

The Environmental Information Regulations 1992 supplement and greatly extend the public's access to environmental information. As stated, the Regulations implement the European Council Directive on the Freedom of Access to Information on the Environment, and were brought into force on 31 December 1992.

Under these Regulations, public bodies with responsibilities for the environment must make environmental information available to any person who requests it.

For the purposes of these Regulations, environmental information is defined widely and includes information on the state of water, air, soil, fauna, flora, land and natural sites and on activities that protect or adversely effect them. Similarly, these Regulations apply to all public authorities and also to private bodies with 'public responsibilities for the environment'.

These bodies are required to respond to a request for information as soon as possible and, in any event, within two months of the request. If the body refuses to provide the information it must give a reason for this refusal, in writing. Possible reasons include that the request was too vague or general or that it was manifestly unreasonable to satisfy the request. The body can also make a 'reasonable' charge for the provision of this information.

However, there are a number of exceptions to the types of information that must be disclosed under these Regulations. These exceptions can be classified into those where refusal is mandatory and those where the body has the discretion to release the information or not.

Mandatory refusal

Grounds for mandatory refusal include: that the information relates to personal information about an individual (and the individual has not consented to its disclosure); that the information was provided voluntarily (and this information could not have been obtained in another way); and that the disclosure of the information would increase the likelihood of damage to the environment – for example, the disclosure of the exact location of nests of rare birds in the UK.

Discretionary refusal

One reason for discretionary refusal is that the information relates to national security, public defence or international relations. Another reason is that the information is, has been or is likely to be the subject of any legal or other proceedings or public inquiry. A third ground allows the body discretion to refuse to release information where it is contained in an unfinished document. However, the most widely used ground for refusing to provide information is that it is commercially confidential.

Problems with the Directive and/or Regulations

In 1996, the House of Lords Select Committee on the European Communities held an inquiry into the effectiveness of the Directive and/or the Regulations and took oral and written evidence from a wide range of bodies, including the Department of the Environment, Friends of the Earth, the Campaign for Freedom of Information and the Confederation of British Industry.

In their report published in November 1996, the House of Lords Select Committee made a number of important criticisms of the Directive, the Regulations and the process for their implementation in the UK. These criticisms are outlined below.

Lack of practical arrangements
The Directive requires member states to make practical arrangements for handling the requests for information. However, the Regulations are silent on these practical arrangements and leave them up to the individual bodies themselves. Given that there are probably over a thousand different bodies subject to the requirements of the Regulations, all essentially making their own arrangements, it is not surprising that this has led to highly variable standards.

Lack of an effective appeals mechanism
If a body refuses to release information, there is no appeal process. If a person wishes to pursue the matter, the only direct recourse is to challenge the decision in the courts. However, a judicial review challenge can be extremely expensive (see Chapter 9 for more on judicial review). Alternatively, a complaint can be made to the parliamentary ombudsman if the case involves a government department. This process can take a considerable time and it only applies to certain public bodies.

In response to these criticisms, the government has stated that it is considering creating a new appeals procedure involving an information commissioner with the power to order the release of information, determine reasonable charges and resolve other disputes.

What constitutes environmental information?
Bodies have shown themselves reluctant to release information on operational or financial matters. For example, the Water Services Association refused Friends of the Earth's request for information on the cost of the water industry's environmental programmes, while English Nature has turned down requests for information on financial aspects of management agreements relating to sites of special scientific interest (SSSI).

The House of Lords recommended that information on operational matters should be released, although the report was less explicit on the issue of financial information.

Which bodies are subject to the Directive and Regulations?
Many privatised utilities, such as the water companies, electricity generators and Railtrack, have claimed that they do not have public responsibilities for the environment and so are not subject to the Regulations. For example, British Nuclear Fuels Ltd turned down a request for information in relation to the operation of the Thorp nuclear reprocessing plant, for this reason. This issue is becoming even more complicated with the increasing range of services (such as waste-disposal operations) contracted out by local authorities.

Until now, the government has been content to leave it up to the bodies themselves to decide whether they are covered. The

House of Lords Select Committee considered that privatised utilities were covered by the Directive and Regulations, and that if they were not then the Directive and Regulations needed to be amended. They were also very critical of the government's hands-off approach and urged them to publish a non-exhaustive list of which bodies they considered were covered.

Charges

The fees charged by bodies vary greatly and can effectively constitute a price barrier to obtaining the information. The most extreme examples include a demand of £365,000 by Ordnance Survey to Friends of the Earth for the supply of digitised map data and a demand of £3,000 by the Department of Transport to Friends of the Earth for traffic count statistics (although this was later reduced to £10).

Some bodies also charge for the time spent assembling the information. For example, Walsall Metropolitan Borough Council demanded that an environmental campaigner pay £40 per hour for time spent extracting landfill gas data from its files. By contrast, other bodies merely charge the copying costs.

The House of Lords has recommended that the Directive's provisions are refined to clarify the meaning of a reasonable charge.

Scope of the exemptions

The House of Lords considered that many of the grounds for exemption have been interpreted very widely. For example, the Health and Safety Executive (HSE) turned down a request from Friends of the Earth for information on safety information relating to major hazard sites – so-called CIMAH (Control of Industrial Major Accident Hazards Regulations) sites – on the basis that the manufacturer had claimed that the information was commercially sensitive and that the HSE did not have the expertise or resources to do anything other than accept the manufacturer's claim. Similarly, the Department of Trade and Industry has cited commercial confidentiality for refusing access to offshore oil exploration licences.

The Ministry of Agriculture, Fisheries and Food refused a request by the Royal Society for the Protection of Birds for information provided by English Nature on applications for bird-killing licences on the grounds that it was volunteered information. English Nature refused to provide the information on the basis that it was an internal communication.

The exemption relating to the disclosure of information that may affect matters that are or have been *sub judice* or under inquiry has also been used to withhold papers submitted to planning inquiries.

The House of Lords Select Committee considered that not only were the Directive and Regulations couched in unnecessarily wide terms but that these had also been interpreted too liberally. They recommended that two new tests should be introduced. The first would require potential harm to be demonstrated by the person seeking confidentiality while the second would allow for the confidentiality to be overridden in cases where public interest so demands.

THE FUTURE

In February 1997, the last government made a cautious response to the House of Lords Select Committee's report. However, the vagueness of the response suggests that many of the report's criticisms and recommendations will not be acted upon.

In addition, the effectiveness of the Directive and its implementation in the member states (including the UK) is due to be reviewed by the European Commission during 1997. This may lead to a strengthening of the Directive in certain areas and so require similar amendments to the Regulations.

While in opposition, the Labour Party promised to expand public access to environmental information and increase the rights of appeal against refusal. Not that they are in power, it is not clear what priority will be given to these promises.

CONCLUSION

The European Union has given campaigners a very significant tool by insisting that the UK, along with the other member states, provides local information regarding developments that affect the local environment. It is up to those affected by developments to use this weapon and it is up to experienced lawyers in the field to challenge instances where information is withheld. By taking these steps we will ensure that Britain becomes a more open society, at least when it comes to environmental issues.

6   Europe

It is not possible to consider the legal rights and duties of individuals and companies in the UK in isolation from those rights and duties imposed by European Union (EU) legislation.

In this chapter, the various institutions that comprise the EU are first described to give the reader a basic understanding. There then follows a description of the various types of EU legislation. Finally, an example of a piece of EU legislation is used to demonstrate the importance within the UK of EU laws and how they may be used by those in the UK seeking to improve the environment.

THE EUROPEAN PARLIAMENT

In Britain we are used to Parliament having the primary decision-making powers. In the European Union the Parliament has been a far weaker body, although its significance is gradually increasing.

The European Parliament consists of 626 members (MEPs) who are directly elected from the member states every five years. The scale of representation of each member state is largely related to the size of its population, so that, for example, Germany elects 99 individual members, France, Italy and the UK 87, Spain 64 and so on, down to Luxembourg with six representatives.

The initial role of the European Parliament was purely consultative but it now shares decision-making powers with the European Council of Ministers in respect of some legislation. Prior to the adoption of any legislative proposal from the

Commission by the Council of Ministers (see below), the Parliament must provide its opinion. In addition, it can amend proposals from the Commission and also the Council of Ministers' preliminary positions.

MEPs form the most direct contact between British citizens and the EU decision-making process and, despite their limited role, it is worth considering contacting or lobbying them on appropriate issues. They may then seek to gain the agreement of their colleagues to support proposed amendments to legislative proposals and therefore the individual EU citizen may have had influence on the law-making process of the EU.

Plenary sessions of the Parliament are held in Strasburg every month, while additional sessions are held in Brussels. The General Secretariat is based in Luxembourg. Meetings of the Parliament are open to the press and public and reports of debates are published in their official journal (impenetrable to all but the cognoscenti or the most diligent reader). In addition, there are European Information Offices in each of the member states that may be contacted directly for information on EU institutions.

THE COUNCIL OF THE EUROPEAN UNION

The Council of the European Union is the most powerful of the EU institutions and consists of a minister from each of the 15 member states, nominated by the individual member state. It adopts community legislation, coordinates the national policies of member states and attempts to resolve conflicts between member states and other EU institutions.

Council decisions are taken either on the basis of qualified majority voting or by unanimity and the presidency rotates among the ministers on a six-monthly basis.

Most decisions are decided by a qualified majority vote with different member states carrying different vote weightings. For example, Germany, France, Italy and the UK hold ten votes each, Spain eight, Belgium, Portugal, Holland and Greece five votes, down to Luxembourg with two. In total, a maximum of 87 votes are possible. The qualified majority is a total of 62 votes, but at least ten member states must support the proposal.

Most environmental proposals are dealt with on the basis of qualified majority voting, although certain proposals, such as regional and social funding that may have significant environmental impacts, are subject to unanimity.

THE EUROPEAN COMMISSION

The European Commission consists of 20 members, two each from the UK, France, Germany, Italy and Spain and one from each of the other member states. The Commission is the executive body of the EU and employs a staff of approximately 15,000. It is divided internally into individual directorates, with the Directorate General XI (DG XI) being responsible for the environment, consumer protection and nuclear safety.

EU legislation starts with the Commission. The usual procedure is for a proposal to be drafted by the Commission which is then submitted to the Council of Ministers. The Commission itself depends upon its individual directorates and panels of experts to be advised on the necessity of legislation and the form that legislation should assume.

The Commission, in theory, should give primacy to the best interests of the EU and its citizens rather than particular interests of individual member states or particular sectors of industry. It should consult interested parties widely and should also respect the principle of subsidiarity, which means that EU action should only be taken when this will be more effective than leaving the action to member states.

The Commission is responsible not only for initiating legislation but also, in some instances, for the implementation of legislation and, in particular circumstances, for the enforcement of EU legislation. For example, if an individual or group is concerned about the apparent non-implementation of EU legislation or the inadequate implementation of EU legislation by a member state, the starting point for any complaint will be the relevant directorate. Numerous complaints have been made by individuals and groups against the

UK's failure to implement or comply with EU legislation in the environmental field over the last number of years.

THE COURT OF JUSTICE OF THE EUROPEAN COMMUNITIES

The European Court of Justice (ECJ) consists of two main institutions. The Court of First Instance was established in 1989 and, as the name suggests, has been given jurisdiction to deal initially with all legal actions brought by individuals or groups bringing complaints against EU institutions or member states.

The second institution, the Court of Justice, is made up of 15 judges, one from each member state, and nine Advocates-General. The President of the Court is elected by the judges and is responsible for presiding over hearings. For particularly important cases the court will sit in full plenary session, but has the power to deal with cases in chambers of either three or five judges.

Actions in the ECJ fall into two broad categories: direct actions and preliminary rulings.

Direct actions may be commenced by the Commission, other European Union Institutions, member states, individuals or groups. They can be cases brought against the Commission, other EU institutions, member states, individuals or groups on a wide variety of bases, but most usually involving direct infringements of EU legislation.

Preliminary rulings occur when a domestic tribunal in a member state is unclear about a point of EU law, and refers a question for clarification to the ECJ. (The Court of First Instance does not have the power to deal with this type of action.) Following a decision of the court the domestic tribunal is bound by EU law to apply the interpretation of the ECJ. There have been numerous instances of this procedure being used from the UK in the last few years.

Since 1989 the Court of Justice has functioned largely as a court of appeal from the Court of First Instance. The Advocates-General are responsible for providing the court with opinions on cases brought before it. It is quite uncommon for the court to depart from an opinion of an Advocate-General.

In terms of procedure, in direct actions the language in which a case is conducted is usually chosen by the applicant, and in cases of preliminary rulings the court will use the language of the national court that referred the matter to the ECJ. Procedures in both courts are predominantly written, and oral argument plays a very small part. Once each party has completed its written (and oral) submissions to the court, the Advocate-General appointed to the case will provide the court with an opinion on the arguments and the interpretation of relevant EU law and the Advocate-General will also recommend a decision for the court to adopt.

Although the procedure outlined above appears to be straight-forward and relatively streamlined, a particular complaint made about the ECJ is the length of time it can take for both direct actions and preliminary rulings to be resolved. It is not uncommon for a period of five years to elapse between the date of an initial complaint to the Commission and a final ruling by the ECJ. This is a particular concern in respect of environmental disputes, when an action is concerned with infringement of environmental legislation. The example of the Directive Concerning the Quality of Bathing Water (known as the Bathing Water Directive) is illustrative of this point (see page 101).

EUROPEAN UNION LEGISLATION

Regulations

Regulations are directly and generally applicable acts of legislation – that is, there is no necessity for individual member states to legislate separately to implement EU regulations provided they are sufficiently precise. Member states are, however, obliged to implement the regulations.

Directives

Directives are binding on all member states and the substance of them allows member states little or no discretion. However, the implementation of directives is invariably left to individual member states. They define the precise means by which a directive is put into effect. In the UK the most common means of implementing EU directives is by means of regulations (not to be confused with EU regulations) made and approved by the relevant secretary of state. Most environmental directives are, therefore, implemented by regulations made by the Secretary of State for the Environment – since May 1997, the Secretary of State for the Environment, Transport and the Regions.

Directives usually set out a timetable to which particular measures must be adopted. In addition, some directives allow for member states to derogate (that is, opt out either from the timetable or from some of the measures set out). As the full implementation of environmental directives invariably involves substantial costs to member states, there is, unsurprisingly, usually a great deal of resistance to full implementation on the part of the UK government. Again, the Bathing Water Directive provides a very good illustration of this point (see page 101).

Direct effect
The issue of whether EU legislation, and directives in particular, are 'of direct effect' is central to EU law. If a law creates or gives rise to legal rights and obligations that can be enforced by individuals or undertakings (that is corporate bodies) then the law will be regarded as being of direct effect.

This concept of 'direct effect' requires the law to be unconditional in the sense that wide discretion is not given to individual member states in defining such laws. Governments and governmental bodies are not allowed to get away with not implementing the requirements of a directive. A third requirement is that for EU law to be directly effective it must be clear and precise enough to form the basis of a legal action, either in a domestic tribunal or in the ECJ.

In the case of *R. v. President of the Board of Trade* ex parte *Duddridge* (1994), the children of residents of an area of north London took the President of the Board of Trade to court for failing to implement the precautionary principle when it came to making regulations controlling the laying of new electric power lines. The residents were concerned that the new line would expose their children to increased risks of contracting childhood leukaemia. The residents argued in the Court of Appeal that the fact the UK government had signed the Maastricht Treaty, incorporating Article 130r, which said that the signatory governments would act in a precautionary way in matters to do with environmental damage, meant that it must take this on board in making regulations.

The Court of Appeal turned down the application, saying there was no obligation on the government until the wording of the requirement was more certain, which meant awaiting the directive derived from the article.

Decisions

Decisions can be addressed to individual member states, to undertakings or to particular individuals. As a general rule decisions are very rare within the field of EU environmental law.

Recommendations and opinions

From time to time the European Council will issue recommendations or opinions. While not binding, these are seen to be of 'persuasive' effect, but have no formal legal means of enforcement.

THE BATHING WATER DIRECTIVE

A good illustration of the way in which EU environmental law impacts upon UK law and the government is provided by the Directive Concerning the Quality of Bathing Water (76/160/EEC), which was approved by the Council of Ministers in 1976. The reason for the agreement of the Directive was concern about the large proportion of contaminated sites for bathing both on the member states' coasts and inland. Of particular concern was the contamination resulting from discharges of raw sewage.

The Directive set out to improve the quality of bathing water by setting minimum standards for compliance and guideline standards that member states should strive to achieve.

The Directive defines bathing water as: 'fresh or sea water in which bathing is explicitly authorised, or is not prohibited and is traditionally practised by a large number of bathers'.

Nineteen parameters are set out in the Directive. These parameters include levels for sewage bacteria (faecal and total coliforms), a viral indicator (enteroviruses), salmonella and a range of other parameters such as phenols, pH and mineral oils.

The first requirement imposed by the Directive upon member states was for them to provide the Commission with a list of designated bathing waters by the end of December 1979.

The guidelines issued from the Department of the Environment decided that unless more than 500 people were in the water together a beach would not constitute a bathing water as defined by the Directive. By applying this definition, the UK presented the Commission with a list totalling 27 bathing waters. Excluded from the list were such resorts as Brighton and Blackpool, which the UK government did not consider to be traditional bathing waters by its own definition. By way of contrast the state of Luxembourg, not previously noted for its bathing facilities, presented the Commission with a list of 34 bathing waters.

The European Commission published a Reasoned Opinion to the effect that, in its view, the UK government was acting illegally by failing to implement the Directive. This view was contested by the UK government but, following threats of legal action from the Commission and financial sanctions, the government reluctantly came up with a list of 364 traditional bathing waters by 1987. By 1995 this list had marginally increased to 425.

Originally, the Directive set a deadline of December 1985 by which member states had to satisfy the minimum water-quality standards as set out in the Directive. In a study from 1987, of the original 27 UK bathing waters designated under the Directive only 56 per cent met the minimum standards.

The statistics published by the National Rivers Authority (NRA) for 1995 show a compliance rate of 89 per cent, but this does not necessarily mean that the quality of these bathing waters fully complies with the terms of the Directive. For example, the figures published by the NRA refer to compliance with 'mandatory coliform standards': that is, the minimum standards required in respect of pollution from sewage bacteria (but not sewage-related viruses).

Some years ago the NRA adopted, as a matter of policy, a sampling regime that would not test for the presence of entero-viruses in bathing waters that had satisfied minimum bacterial standards the previous year. Many experts are quite convinced that, if waters were to continue to be tested for the enterovirus parameter, the compliance rate would be significantly lowered.

As regards the more stringent guideline standards set out in the Directive, of the 425 bathing waters designated in England and Wales only 18 satisfied the guideline standard in 1995.

The UK position as regards inland bathing waters is even more stark. In 1980 France had designated 1,362 inland bathing waters. At the time of writing the UK has failed to designate a single one. Once again, the reasons for non-designation are not difficult to find. In 1991 the NRA stated in its annual report:

> Many inland UK sites if identified, especially those on rivers, would not at present pass the directive's microbiological standards, because as well as point source inputs of sewage effluent, there are many diffuse sources of bacteria which enter rivers from farms, bird flocks and so on. It would be very difficult to treat all the possible diffuse sources of bacteria to the water, and hence difficult to guarantee that a site would pass the Directive standards.

The approach of the UK is thus neatly summarised. If compliance with the Directive will involve expenditure, then better to avoid compliance than incur increased expenditure for the water industry. It seems that British farms and birds produce a particular type of sewage which is rather more difficult to treat than sewage produced by French farms and birds.

The Commission remains unimpressed by Britain's record of designation and implementation of the Bathing Water Directive, and the UK was taken to the European Court of Justice by the

Commission. The ECJ ruled in July 1993 of that year that Britain was in breach of the Directive in relation to bathing water quality at both Blackpool and Southport (which had been selected by way of examples).

It is quite clear that the driving force behind the UK government's numerous attempts to circumvent the terms of the Bathing Water Directive are almost exclusively financial considerations. With the change of government in May 1997, it may be expected that some different considerations will be employed. It is bad enough that local authorities had to incur substantial expenditure in bringing sewage discharges up to minimum standards in the mid-1980s, even worse now to present newly privatised water companies with massive liabilities for costs of compliance.

As a postscript it is worth mentioning a more recent attempt by the UK government to avoid the costs of complying with another EU directive, once again in the field of sewage treatment.

The Urban Waste Water Treatment Directive 91/271/EEC requires particular levels of sewage treatment for discharges from particular populations into estuarine and coastal environments. The purpose of the Directive is to ensure high levels of treatment for sewage from large towns discharging to sea.

A minimum of secondary treatment is required for discharges from urban centres into estuaries. Less stringent treatment is only possible in respect of discharges to coastal waters and estuaries that are in 'less sensitive areas', provided such discharges do not adversely affect the environment.

It is the responsibility of the Secretary of State for the Environment, Transport and the Regions to identify less sensitive areas. In November 1994 the Secretary of State for the Environment designated as coastal waters an area of the Humber Estuary up to the Humber Road Bridge (a distance of some 20 miles from the coast) and the Severn Estuary up to the Severn Road Bridge. The intent was clear! It would mean that sewage discharges from the conurbation of Kingston-upon-Hull would require only primary as opposed to secondary treatment. In remarkable contrast to the position adopted by the Secretary of State for the Environment in the 1980s, where Blackpool was not deemed to be a traditional bathing area, in 1994 both Kingston-upon-Hull and Bristol became coastal resorts.

This decision was challenged by way of judicial review in the High Court in 1996 and, unsurprisingly, the Secretary of State was asked by the court to think again.

THE LAPPEL BANK CASE

On 2 April 1979, the Council of Ministers approved Council Directive, 79/409/EEC, on the Conservation of Wild Birds, usually referred to as 'the Birds Directive'. The Directive requires individual member states to designate Special Protection Areas (SPAs) for the protection of wild birds. The Birds Directive was seen to be necessary to maintain bird populations and habitats for wild birds, which had come under increasing pressure, particularly from industrial development through the 1960s and 1970s.

On 21 May 1992, the Council Ministers approved a further Directive, 92/43/EEC, on the Conservation of the Natural Habitats of Wild Fauna and Flora, usually referred to as 'the Habitats Directive'. As with the Birds Directive, the Habitats Directive was a response to increasing pressure on the natural environment and was seen as being necessary to prevent the further deterioration of natural habitats. Certain articles of the Habitats Directive replaced some of those in the Birds Directive, the most important being in respect of those considerations that member states should take into account when designating Special Protection Areas.

In December 1993, the UK Secretary of State for the Environment decided to designate the Medway Estuary and Marshes as a Special Protection Area. This was on the basis that the estuary and marshes form an area of wetland of international importance and are used as breeding and wintering areas for a wide variety of birds. However, the Secretary of State decided not to include an area of roughly 22 hectares, known as Lappel Bank. Although forming an integral part of the Medway Estuary and Marshes, Lappel Bank adjoins the port of Sheerness to the north. In 1989, planning permission was granted to expand the port of Sheerness. In particular, the area of Lappel Bank was earmarked to provide an enormous car and cargo park for the port.

The Secretary of State decided to exclude Lappel Bank from the Medway SPA on the basis that local and national economic interests outweighed the nature conservation interests at Lappel Bank.

This decision was challenged by the Royal Society for the Protection of Birds (RSPB) early in 1994. The RSPB took out a judicial review in the Divisional Court and this court upheld the decision of the Secretary of State for the Environment.

The RSPB appealed this decision to the Court of Appeal but the appeal was dismissed in August 1994. In addition, the Court of Appeal refused to refer the matter to the European Court of Justice by way of Article 177, saying that the directives in question were sufficiently clear and that the court did not need to ask the ECJ for guidance. In addition, the Court of Appeal refused to grant the RSPB an injunction, which would temporarily halt the construction work at Lappel Bank until the legal case was finished.

The RSPB then appealed against the decision of the Court of Appeal to the House of Lords. In February 1995, the House of Lords, while refusing the appeal, decided to refer the question of the interpretation of the two directives to the ECJ. Once again a temporary injunction to halt development work was refused.

Not until July 1996 did the ECJ rule on the question of whether economic considerations could take priority over conservation considerations in respect of the Birds and Habitats Directives.

The ECJ held that the UK government had acted illegally by excluding Lappel Bank from the Medway Special Protection Area and that economic interests could not override the conservation requirements of the Birds Directive. In fact, the ECJ held that a member state was not authorised to take account of economic requirements when designating Special Protection Areas.

The Lappel Bank case provides, in a nutshell, a perfect illustration of the relationship between the UK and European Union courts. It also provides a graphic illustration of the limits of EU law when faced with an intransigent domestic judicial system. Lappel Bank is today a car park.

Part II

ACTIONS TO STOP NEW DEVELOPMENTS

7 Planning

The planning system is very structured and environmental campaigners may well decide, for reasons that will be described in detail below, to manage and run opposition to a development or plan themselves. It is vital, therefore, that campaigners understand the system and the opportunities to intervene. This chapter is consequently more detailed than the other chapters in this book.

WHAT IS TOWN AND COUNTRY PLANNING?

Town and country planning is the statutory system introduced in Britain after the Second World War to regulate land use and development in the public interest. Its aim is to strike a balance between various conflicting interests, to ensure that the needs of different users of land are all taken into account. Of particular importance for the environment is that the system is designed to make a decision about the acceptability of particular form of land use *before* it happens, rather than after. Furthermore, because it embodies extensive public consultation procedures and relatively transparent decision-making processes, it lends itself well to a campaigning approach.

The planning system comprises two main parts. The first, forward planning, involves the making of development plans. In this way local authorities set out how they think land use and development should take place in the future. The second, development control, provides a regulatory framework in which proposals for development are controlled, in the context of the development plan.

Planning authorities

There is a range of bodies with a part to play in the planning system, reflecting decision making by the 'local planning authority' within a national framework supervised by the government. In areas where there is a unitary local authority, including all metropolitan districts and London boroughs, it is the unitary authority that is the local planning authority. In all other areas, where there are two tiers of local government, the planning functions are shared between them. In forward planning, county councils are responsible for producing structure plans and local plans for waste and minerals whereas district or borough councils are responsible for producing area-based local plans. Unitary local authorities must make unitary development plans (UDPs).

Development control is primarily a district function, except for waste and minerals matters which are determined by the county planning authority. Special provisions apply in National Parks and the Broads and in Urban Development Corporation areas.

The Secretaries of State

The role of the Secretary of State for the Environment, Transport and the Regions (in England), or the Secretary of State for Wales (in Wales), is to ensure that local decision making takes place in a national framework with some degree of consistency. The Secretaries of State issue guidance to local planning authorities on how to carry out their functions and on how they will consider appeals against local authority decisions. Local authorities must have regard to this guidance in planning matters. The guidance appears in the form of planning policy guidance notes or PPGs (currently numbered 1 to 24, e.g. PPG 23 on planning and pollution control), regional planning guidance notes (RPGs), minerals planning guidance notes (MPGs) and circulars from the Department of the Environment, Transport and the Regions, and the Welsh Office. Not all PPGs apply in Wales, and there are a number of Technical Advice Notes (TANs) which only apply in Wales.

DEVELOPMENT PLANS

The importance of development plans for campaigners

Because development plans play such an important part in all development control decisions, it is crucial that environmental campaigners intervene early on in the plan-making process. The primary point is that the relevant authorities are legally obliged to take planning decisions in accordance with the development plan, unless there are good reasons for this not being the case and, therefore, it is vital that campaigners start with this process. Otherwise, when an important planning application is being considered, it is much more difficult to resist an unacceptable development. For example, when a local group wants to resist an application for the development for housing of an informal open space in their area, which has both local amenity and nature conservation importance, if the local plan or UDP has already allocated the site for housing use, it will be very difficult to overturn it. It will be necessary to show that there are reasons why the use should not be implemented. If, on the other hand, objectors successfully intervene when the local plan or UDP is in the process of preparation, it will not be necessary to overcome a presumption in favour of developing the site.

It is, however, understandable that local groups often do not intervene in the planning process until too late. Although plan making will be publicised, it will be in general terms. Unless a specific enquiry or investigation has been made by local campaigners it is quite possible that an unacceptable policy proposal relating to a site of particular concern will not come to local attention. Because there is no immediate plan to develop the site, local attention is not focused as effectively as it would be if there were a planning application for the site.

Types of plan

What constitutes the development plan will depend on the type of area:

- In a unitary authority area it will be the unitary development plan (UDP).

- In any other area it will consist of the following documents: the county structure plan, the district or borough local plan and the waste and mineral local plans produced by the county council.

Structure plans
The structure plan must contain the statement of the county planning authority's general policies and proposals regarding development and other land-use issues in its area. These policies must deal with conservation of natural beauty and amenity of land, the improvement of the physical environment and the management of traffic. It is intended to be a strategic document indicating the scale and general location of types of uses to be restricted. It is produced in the form of a written statement with diagrams that are also part of the plan and with an explanatory memorandum that gives reasons for every policy. The structure plan must have regard to environmental issues.

Local plans
A local plan will contain a written statement of the authority's detailed policies for the development of land use in its area. The plan must be in general conformity with the structure plan and largely fills in the details within the framework set out there.

For 'minerals local plans', county planning authorities are required to produce a written statement of their policies on the winning and working of minerals and on the deposit of mineral waste in the area. 'Waste local plans' must contain the policies concerning the deposit of waste or refuse and must have regard to waste-disposal plans drawn up by waste regulation authorities (now the Environment Agency). These two types of local plan are sometimes combined as 'minerals and waste local plans'.

Unitary development plans
In unitary authority areas the unitary development plan (UDP) constitutes the only plan for the area. It is in two parts which

broadly resemble a structure plan and an area local plan. The same considerations apply as are set out in those sections. Policies relating to waste and minerals will also be included within the UDP.

Development plan-making procedure

Development plan making is a continuous process. Once a plan has been adopted the local planning authority must keep it under review and bring forward changes periodically. Concerned individuals or groups need, therefore, to find out the stage that each of the relevant plans has reached and make an appropriate intervention on the issues that concern them.

The first stage of plan production or revision involves the planning authority in reviewing relevant factors in its area. Consultation must then be carried out with a range of bodies on its proposals before a draft plan or plan revisions are put on deposit. Campaigners should ensure that the planning authority is aware that they want to be continuously consulted from this stage.

The planning authority must then consider any representations made and draw up a further draft, known as the deposit draft, with a memorandum explaining its conclusions. This more formal draft is then put on deposit and must be advertised and made available for public inspection and consultation. Those wishing to make representations will be given a period of six weeks in which to do so. It is very important that objectors either obtain a copy of the deposit draft or inspect one at the planning department or local library. The whole plan must be considered, not just proposals for particular sites. Policies in the plan area can also have a great effect on planning applications at a later stage. Objections can take the form of opposition to policies and proposals included in the draft and objections to omissions from the plan. Representations can be made in support of acceptable policies and proposals. These representations must be submitted in writing within the time period.

The planning authority must then consider the representations and objections and prepare a statement on each one giving its response and reasons for it. In the case of a unitary development plan or local plan, if there are any outstanding objections at this stage the planning authority must hold a public local inquiry. All objectors will have the right to appear in support of their objections before the inquiry, although they can make written submissions instead if they wish. In a case of a structure plan, an examination in public (EIP) will be held into those matters the authority considers should be examined. The EIP is a much more limited form of hearing than a public inquiry and only considers selected issues. No person has the right to be heard at an EIP except at the invitation of the panel chair or planning authority.

Before the inquiry or EIP the planning authority will consider the objections made and may attempt to negotiate an acceptable compromise. The role of objectors is extremely important in determining the final shape of the plan. It must be remembered that the main participants in the process tend to be developers, not only those with interest in special local sites, but also national organisations, such as British Telecom and British Rail, which monitor policies that affect their interests over the whole country.

Development plan inquiries and EIPs have a number of differences from other planning inquiries. The section in this chapter on planning inquiries (page 137 below) is of relevance but there are a number of differences set out here. In these cases, the inspector or panel is charged with reporting to the local planning authority to help it deal with objections to the deposit draft. This is different from other inquiries which have the function of either determining appeals or reporting to the Secretary of State.

Development plan inquiries and EIPs are like a series of different inquiries with different parties and tend to be more or less formal depending on the parties involved at the particular time. Although the local planning authority is usually represented by a barrister or at least a solicitor, many of the sessions are much less formal than a public inquiry into a planning appeal. As the inquiry or EIP frequently lasts for a period of months, the local planning authority will appoint an independent programme officer to assist the inspector or panel in running it. The programme officer is the liaison point for objectors and it is important to make contact with this person at an early stage.

A pre-inquiry meeting will normally be arranged by the inspector or chair to consider the programme, timetable and procedure, and it is important for objectors to attend this meeting. Since, unlike other planning appeals, no procedure is laid down for development plan hearings, this is a matter for the inspector or chair to decide. It is therefore important for objectors to be aware of the approach that will be taken by the inspector or the chair.

Proofs of evidence are usually exchanged before the inquiry or EIP. In the case of shorter hearings, these are usually required to be provided by objectors at least six weeks before the start of the hearing, with local authority proofs in response to be produced three weeks before the hearing. In longer hearings, a rolling or staggered programme of proof production may be adopted. The timetable should allow adequate time for objectors to consider the local authority proofs before they appear at the hearing. Because the local planning authority has a massive job to produce proofs dealing with all the objections and to appear at the inquiry, often at the same time, it is very common for them to fall behind the timetable. Some flexibility is of course necessary; however, in cases where a failure to comply with the timetable is causing particular difficulties for objectors, this should be raised with the programme officer, who will pass it on to the inspector or chair.

After the inquiry or EIP, the inspector or panel prepares a report with reasoned recommendations, which can take a matter of many months or, sometimes, more than a year, and this is sent to the local planning authority. The authority then considers the report and decides what amendments, if any, it proposes to make in the light of it. The report and a statement of the planning authority responses are made available by the authority with any modifications it proposes to make to the plan. If any of these modifications materially affects the contents of the plan, the authority must publish a notice to this effect and invite further objections and representations within a period of six weeks.

If there are no material modifications proposed, the authority will give notice of intention to adopt the plan. Adoption can take place after 28 days if no modifications were recommended by the report – otherwise objections to the failure to adopt the report recommendations must be made during the period of six weeks.

Any objection at this stage must relate to the modification proposed or failure to make a modification proposed in the report. The authority must consider all objections and decide whether to reopen the hearing. If a new inquiry or reopened EIP is held, the planning authority must go through the same procedure regarding advertising, etc., as before. A copy of the adopted plan is an essential item for environmental campaigners intending to play an active part in the planning process.

There are limited grounds on which the adoption of a development plan can be challenged in the High Court.

DEVELOPMENT CONTROL

What is controlled?

The development control regime allows local planning authorities to regulate building or other physical alterations to land or changes of its use. This is done primarily by means of the grant or refusal of planning permission. If alterations are carried out without permission the local planning authority has the power to take enforcement action ultimately culminating in criminal prosecution. Breach of planning control, in itself, is not a criminal offence.

The legal framework of development control is very complex and the detail of it cannot all be covered here. However it is necessary to have an understanding of how the broad framework works, in order to be able to intervene effectively in the process. In general terms any development of land requires planning permission.

The definition of development has two branches:

- the carrying out of building, engineering, mining or other *operations* in, on, over or under the land or,

- the making of any material change in the *use* of any buildings or other land.

A distinction between these two branches is not always as clear cut as it seems. There are many exclusions to this rule, where planning permission is not required.

Certain changes of use set out in the Town and Country Planning (Use Classes) Order 1987 do not require permission. Uses are grouped into classes – for example, business use, which includes office, research and development, and light industry. Changes within such a group do not require permission. Other changes between groups do not require permission, such as from general industry to business. A range of other activities that would otherwise be considered to be development are effectively exempted by the grant of deemed planning permission under the Town and Country Planning (General Permitted Development) Order 1995 and a number of special development orders. Exemption is also granted in specific geographical areas by the creation of enterprise zone schemes and simplified planning zones, and the Crown is totally exempt from planning control.

Planning applications and permissions

Every planning application must be made on a form provided by the local planning authority and it must be accompanied by a site plan identifying the application land and other plans describing the development proposed. The authority can require further information to clarify the development proposed.

Developers can choose to seek outline planning permission. Such a permission would determine that the development under consideration is acceptable in principle. In such cases all that is supplied is a site plan and a brief description of the development proposed. If permission is granted it would be made subject to a condition that reserved matters such as the siting, design, external appearance, means of access and landscaping are all subsequently approved by the authority.

All planning applications are made to the unitary authority or non-unitary areas to the relevant district council (except for applications relating to county matters, which are made direct to the county council). A fee is payable for all planning applications. Each application must be accompanied by a certificate that confirms that the applicant owns all of the application land or has notified or attempted to notify the owners of the land of the proposal. The planning authority must then publicise the application either by using a site notice or by notification of neighbours. In addition, advertisements must be published in a local newspaper for those developments that are of a larger scale or may have greater impact on the environment.

Environmental campaigners concerned about specific sites or specific types of development can require the local planning authority to advise them if any relevant applications are made. Local planning authorities must also consult a range of statutory bodies depending on the nature and/or location of the proposed development. Any representations made either by statutory bodies or by anybody else must be taken into account by the authority in determining the application.

Call-in powers
The Secretary of State retains the power to monitor refusals of planning permission by planning authorities using the appeals process. If permission is refused for a development the Secretary of State would wish to grant, he/she can do so on appeal. However, since there is no appeal against the grant of planning permission, the Secretary of State has the power to 'call in' applications, which local planning authorities intend to approve, for his/her own decision. In these cases, the Secretary of State puts him/herself in the position of the local planning authority and determines the application him/herself.

To ensure that he/she is aware of applications that might warrant a call-in, local planning authorities have a duty to notify the Secretary of State of certain types of applications that they do not propose to refuse. These are large-scale applications that do not accord with the development plan and the planning authority's own developments, or those that are proposed for the local planning authority's own land, or those that would, if permitted, significantly prejudice the implementation of development plan policies and proposals. In those cases the local planning authority cannot determine the application for 21 days following notification to the Secretary of State to allow the application to be called in.

The call-in powers are rarely used and are only usually put into operation in cases where the Secretary of State considers the application to be of more than local importance. If an application is called in, determination by the Secretary of State will follow a public inquiry. This determination has to be made in accordance with the development plan unless material considerations dictate otherwise.

Material considerations
Planning applications must be determined by planning authorities having regard to the provisions of the development plan and any other material considerations. It is clearly of great importance for campaigners to know what might or might not be a material consideration. There is no clear dividing line, but the following lists may help.

Items that are always material considerations:

• government policy;

• relevant development plan policies;

• an environmental impact statement;

• the planning history of the site – for example, existing planning permissions, previous refusals, appeals;

• representations by statutory consultees and others in response to consultation;

• the effect on a conservation area or listed building;

• designated status or site or surroundings: for example, area of outstanding natural beauty, site of special scientific interest.

Items that are sometimes material considerations:

• third party or neighbour interests – these are problematic: whereas the loss of a view has been held not to be material, the overlooking of an existing property by new development may be material;

• personal circumstances of the applicant or neighbours – not generally material, but sometimes the specific needs of the applicant or a relative, a sensitive neighbouring use or the need for affordable housing may be material;

- 'planning gain' – may be material.

Items that are never material considerations:

- matters that are not related to the use of land – for example, ownership or impact on value of neighbouring property;

- overtly political factors, such as the policy stance of a local political party or group.

Determination of applications
The local planning authority must determine a valid application within eight weeks of receipt (or 16 weeks if an environmental assessment is carried out). It may grant permission unconditionally, or subject to conditions, or it may refuse permission. If conditions are imposed or permission is refused, the authority must give full reasons in the decision notice.

It will be necessary for any person wishing to intervene in the decision-making process of a planning application to ascertain who will actually be making the decision within the authority. Local authorities have the power to delegate such decisions to a committee of elected councillors or to an officer of the council. Determination of uncontroversial applications that have not attracted objections is frequently delegated to the chief planning officer. Otherwise decisions are usually made by the planning or development control committee. This will depend on the standing orders or rules made by each council to regulate its own practice. Particularly important planning decisions may be made by the full council.

There is no principle that gives either an applicant or an objector a right to be heard before an application is determined, but the general rules of natural justice require that all representations that have been made to the authority must be considered before the decision is taken. In the interests of fairness it is also necessary that, if those in favour of the development are allowed to address the planning committee, the same facility should be granted to the objectors.

Internal local authority organisation may mean that the majority group of councillors has already considered the application at a political party meeting before the planning committee. This has been held to be lawful, provided that those members take account of a detailed and balanced report prepared for the committee by council officers.

Local planning authority applications
Particular difficulties are experienced by objectors when the planning application under consideration is one that relates to land owned by the planning authority or where the authority is involved in a development deal. In such circumstances, it may be particularly difficult to be sure that the application has been given proper impartial consideration. Unfortunately these concerns can only be expressed to the council, and ultimately, if permission is granted, the only remedy is by way of judicial review.

Planning conditions
Conditions imposed on planning permissions must reasonably relate to the permitted development. This means that the council cannot impose restrictions on an earlier element of development. In addition, every planning permission must be subject to a condition that development shall be commenced within five years or such other period imposed by the planning authority. In the case of outline permissions, applications for approval of reserved matters must be made within three years of the grant of outline permission.

Planning obligations
Frequently objectors will hear that a planning obligation, planning agreement or 'Section 106 agreement' has to be prepared before permission will be granted. These planning obligations may be used to secure benefits or restrict development to overcome planning objections. They are frequently used by developers to encourage planning authorities to accept development which might otherwise be unacceptable. The law relating to planning obligations is particularly complicated and if objectors have concerns about the nature of an obligation being entered into they should seek legal advice.

Intervention against planning applications

There are several steps that objectors can take to intervene against a planning application.

- Find out what is proposed and the timescale for the decision. It is possible to inspect the planning application at the council's planning offices. Also ask when the decision on the application is likely to be made.

- If you are unhappy with proposal, contact the planning officer at the council allocated to deal with the case. Find out his/her initial view. Ask about relevant development plan policies. Ask when the decision will be made and by whom – by a committee or by individual officers. Ask the names of the councillors on the planning committee making the decision.

- If appropriate, contact the council's environmental health or highways departments and statutory consultees such as the Environment Agency.

- Contact other bodies such as local branches of the Council for the Protection of Rural England, Friends of the Earth, or Campaign for the Protection of Rural Wales.

- Consider contacting the developer to clarify the position.

- Clarify the planning basis for any objection.

- Make representations to the council. Write to the chair of the council planning committee explaining the objection and giving reasons. Lobby councillors at their surgery and/or ask for a meeting. Write to all councillors on the committee that is making the decision. Ask for a deputation to the decision-making committee meeting.

- Start campaigning. Encourage other people to write to the council and/or statutory consultees. Ask the MP to write to the council. Contact the press.

- In appropriate cases write to the Department of the Environment, Transport and the Regions at the government regional office to ask for the application to be called in.

- Get the committee report that includes the officers' recommended decision, which must be available three clear days before the meeting.

Appeals to the Secretary of State

An applicant for planning permission has a right of appeal to the Secretary of State against a refusal of planning permission or the grant of permission subject to conditions. If a decision has not been issued within eight weeks of the application being lodged (or 16 weeks in the case of a planning application where an environmental assessment is required) and unless the developer has agreed to extend the period, an appeal can be made to the Secretary of State against non-determination of the application. Notice of appeal must be lodged within six months of the date of decision or, in the case of non-determination, within six months of the expiry of the period for determination. The notice of appeal must be sent to the Secretary of State with a copy to the local planning authority. Only the applicant for planning permission can appeal. The Secretary of State reconsiders all the evidence and arrives at a fresh decision.

Enforcement action

If a development is carried out in breach of planning control, the local planning authority has a range of powers to force developers to comply. Broadly, the system operates as follows. In response to a breach of planning control, the local planning authority may serve an enforcement notice against which the owner or occupier of the land has a right of appeal. The enforcement notice does not come into force pending the determination of an appeal. The notice only comes into effect if the appeal is unsuccessful and following a period for compliance. Breach of an enforcement notice is a criminal offence.

In appropriate cases the local planning authority may take other action. If they consider it necessary to stop the breach of planning control pending the outcome of an appeal, they may serve a stop notice. They can serve a breach of condition notice, failure to comply with which is a criminal offence, or they can seek an injunction from the High Court. If they require further information about the breach of planning control they can serve a planning contravention notice.

Enforcement notice
A local planning authority may issue an enforcement notice where it appears to them that:

(a) there has been a breach of planning control; and

(b) it is expedient to issue the notice, having regarding to the provisions of the development plan and to any other material considerations.

A breach of planning control comprises either:

• the carrying out of development without the required planning permission; or

• failing to comply with any condition or limitation subject to which planning permission has been granted.

Immunity from enforcement
Enforcement action must be taken within four years of a breach of planning control involving building, engineering, mining or other operations without permission. For all other breaches of planning control, enforcement action must be taken within ten years of the date of the breach. In order to establish whether an existing use or other form of development is lawful and, therefore, immune from enforcement, application can be made to the local planning authority for a certificate of lawfulness of existing use or development (CLEUD). The grant of a CLEUD has the same affect as a planning permission. It is also possible to apply for a certificate of

lawfulness of proposed use or development (CLOPUD). Such a certificate confirms that a proposed use or development would be lawful, without the requirement for further grant of planning permission.

Issue and service of an enforcement notice
The courts have repeatedly confirmed that local planning authorities have wide discretion in determining whether to serve an enforcement notice. As a result, environmental campaigners concerned about the impact of unlawful development sometimes have a difficult task in persuading planning authorities to act. Action that can be taken is discussed below.

Once a notice has been issued by the local planning authority it must be served upon the owner and occupier of the land and any person with a material interest in the land. It must specify the matters that appear to the local planning authority to constitute the breach of planning control and must specify the steps required by the authority to be taken or activities required to be ceased to remedy the breach or any injury to amenity caused by it. It must also specify the date on which the notice is to take effect, which must be at least 28 days after the date of service of the notice. It must also specify the period thereafter for the steps required to have been taken to comply with the notice. Details of the right of appeal to the Secretary of State and procedure for making an appeal must accompany the notice.

Appeal against the enforcement notice
Any person with an interest in the land or who occupies it may appeal against the notice to the Secretary of State. Written notice of appeal must be given to the Secretary of State before the date on which the notice is to take effect.

There are a number of grounds of appeal. Broadly they are:

- there has been no breach of planning control that can be enforced against;

- the action required to remedy the breach is excessive;

- the timescale for complying is too short;

- planning permission should be granted or the condition or limitation should be lifted.

Where an appeal is made against an enforcement notice, the effect of the notice is suspended pending determination of the appeal by the Secretary of State.

Breach of the terms of an enforcement notice that has taken effect is a criminal offence. The owner of the land is guilty of an offence unless he/she can show that he/she did, within reason, everything to secure compliance with the notice. Any other people who have control of or an interest in the land commit an offence if they carry on an activity required by the notice to cease, or cause or permit such an activity to be carried on. In a magistrates' court the maximum penalty for non-compliance with an enforcement notice is a fine of £20,000. In the Crown Court the fine is unlimited.

In addition to the powers of prosecution, the local planning authority may, once the compliance period has expired, and where works have not been carried out, go onto the land and take the steps required in the notice. They can then seek to recover the costs from the owner.

Stop notice

The stop notice provides a powerful tool for local planning authorities facing serious breach of planning control. In appropriate circumstances this power can be exercised almost immediately. Breach of the stop notice is a criminal offence. Because in very limited circumstances compensation may be payable if the notice was wrongly served, many local planning authorities are unwilling even to contemplate using it. Provided the local authority is careful in collecting its information and drafting and serving the notice there is no real risk of an award of compensation, but the campaigner may find it an uphill task to try to convince the local authority of this.

The stop notice is an interim remedy to be used at the time of service of an enforcement notice or after. It can bring any or all of the requirements of the enforcement notice into effect within a very short time. It takes away the right of the recipient of an enforcement notice to suspend the notice by means of an appeal and to carry on with the breach of planning control until the ultimate determination of the appeal, which can be a matter of many months. Authorities are advised by the Department of the Environment, Transport and the Regions that they should only stop activities that it is essential to stop 'to safeguard amenity or public safety in the neighbourhood; or to prevent serious or irreversible harm to the environment and the surrounding area'. There is no right of appeal against a stop notice. Contravention or causing or permitting the contravention of a stop notice is an offence carrying a penalty of up to £20,000 fine in a magistrates' court and an unlimited fine in the Crown Court.

Given that the issue of compensation is of particular concern to local authorities, campaigners attempting to persuade local planning authorities to use a stop notice will usually need to apply a good deal of pressure for those concerns to be overcome. Clearly if the local planning authority has determined not to serve a stop notice on the basis of a misunderstanding of the law regarding compensation – that is, that they think they will be liable to pay it – then this is potentially a ground for judicial review.

The general principle is that a person who has an interest in land, or is an occupier when the stop notice is first served, is entitled to compensation from the council for any loss or damage directly attributable to the prohibition in the notice in a range of circumstances. In general terms, compensation is payable when the stop notice is withdrawn or the enforcement notice is withdrawn or quashed or varied, lifting the prohibition on the relevant activity, except where it is as a result of the grant of planning permission, either by the local planning authority or on appeal. Compensation is not payable in respect of the prohibition of any activity that, at the time when the notice was in force, was in breach of planning control. Furthermore, compensation is now not payable in cases where a planning contravention notice has been served on the claimant and the loss could have been avoided had information been supplied by the claimant or had he/she otherwise cooperated with the local planning authority when responding to the notice.

Breach of condition notice

If there is a breach of a condition on a planning permission, the local planning authority may serve a breach of condition notice on any person who is carrying out or has carried out the development or any person having control on the land. These notices apply solely to conditions regulating the use of land. The notice must specify the steps required to be taken or the activities required to be ceased to comply with the condition. There is no right of appeal against the notice. Failure to comply with a breach of condition notice within a period specified in the notice (not less than 28 days) is a criminal offence. The only defences are where the defendants can prove that they took all reasonable measures to secure compliance with the condition or where they can show they are no longer in control of the land.

Injunctions

The local planning authority has a power, where it considers it necessary or expedient to restrain any actual or apprehended breach of planning control, to seek an injunction. It can apply to the High Court or county court for an injunction whether or not it has exercised its other enforcement powers. The penalty for breach of an injunction may be a more effective sanction since, in appropriate cases, a person in breach risks imprisonment for contempt of court.

Planning contravention notice

Where a local planning authority considers there may have been a breach of planning control, a planning contravention notice can be served on an owner or occupier of land or a person carrying out operations or using the land. The notice can require the provision of details of operations on the land, the use of the land or other activities and any limitations that have been imposed on planning permissions granted in the past. Failure to comply with a notice without reasonable excuse or the provision of false or misleading information is a criminal offence. The local planning authority can arrange a meeting to provide a formal opportunity for the owner or occupier to discuss the breach. In addition, the local planning authority has powers of entry onto any land in order to collect information, and if admission has been refused or is expected to be refused, a warrant can be sought from a magistrate to secure entry.

Intervention to urge enforcement action
There are several steps that objectors can take to put pressure on the planning authority to take enforcement action:

- Make an appointment to inspect the planning register. Ask to see the files held in the planning department dealing with the site in question. (See Chapter 5 for more details on access to information.)

- Attempt to particularise the nature of the objectionable activity or other development – for example, long hours of work, inappropriate use of land, new building, etc.

- Clarify from the planning register and files exactly what development has been authorised by grant of planning permission. Particularly examine the conditions imposed, if planning permission has been granted.

- If development appears to be in breach of planning control, write formally to the Chief Planning Officer specifying the nature of the development and how it is considered to be in breach of planning control. If appropriate, seek to meet the enforcement officer on site.

- Chase a response, as there is frequently a delay. Threaten to refer to the ombudsman.

- Lobby councillors on the planning committee and local councillors.

- Consider judicial review (see Chapter 9), and take legal advice. Consider other forms of legal action (see Chapter 10).

- Contact the local MP. Contact the press but be careful about the truth of the allegations made.

Special cases

We have already considered the main areas of planning control in detail. There are, however, a number of related controls that arise frequently in conjunction with other planning matters and a summary of some of these provisions may be helpful.

Listed buildings and conservation areas

Special protection is provided under the Planning (Listed Buildings and Conservation Areas) Act 1990 for buildings of special architectural or historical interest and conservation areas. Buildings of special architectural or historical interest are 'listed' by the Secretary of State for Culture, Media and Sport (formerly Secretary of State for National Heritage), in England, and the Secretary of State for Wales, in Wales. Once listed, a building and its setting are protected against inappropriate development. It is a criminal offence to demolish, alter or extend such a building in a way that would affect its character without listed building consent. Planning permission will also not be granted for development that would detrimentally affect the setting of such a building.

Local planning authorities have a duty to designate any part of their area that is of special architectural or historical interest, the character or appearance of which it is desirable to preserve or enhance, as a conservation area. Once designated, in considering planning applications in a conservation area, the local planning authority must pay special attention to the desirability of preserving or enhancing the area's character or appearance.

Urban development corporations

In a number of run-down inner city areas, urban development corporations have been established under the Local Government, Planning and Land Act 1980 to secure the regeneration of those areas. Such bodies have been established in Merseyside, London Docklands, Central Manchester, the Black Country, Teesside, Tyne and Wear, Cardiff Bay, Leeds, Bristol and Sheffield. The corporations are given wide powers, including the acquisition and disposal of land, and development control powers have been

transferred to them. They become the local development control
authority, although development plan-making functions remain
with the usual appropriate local planning authority.

National parks and areas of outstanding natural beauty
Ten national parks have been established across England and
Wales. Recent changes provide for the establishment of national
park authorities which have many of the functions of local
authorities. In planning terms, national park authorities have the
duty to preserve or enhance the natural beauty of the area.
Additional areas known as areas of outstanding natural beauty are
designated under the National Parks and Access to the Countryside Act 1949. In these areas, although planning functions remain
with the local authority they have similar duties to preserve or
enhance the beauty of the area as in national parks.

ENVIRONMENTAL ASSESSMENT

The requirement for assessment of effects of certain public and
private projects on the environment, imposed by European
Directive 85/337/EEC, has been grafted onto the planning process
in the UK in most cases. The government argued that environmental impact is and always has been a material consideration
taken into account in determining planning applications. Whether
or not this was true, it is now included as a formal requirement in
major development with potentially significant environmental
effects. The aim of the Directive was to ensure that, for projects
with a large environmental impact, an assessment of the impact of
the development would be taken into account, with other factors,
in determining whether it should proceed. The prime consideration behind the policy development was that it is appropriate to
try to prevent the impact of pollution or nuisance rather than to
rely on dealing with the effects once the project is operating.
    Environmental assessment allows for the environmental impact
of a project to be considered and, in the light of the process, for
the project to be modified to minimise its effects. The assessment
is a process not a document, which is incorporated into the

regulatory authority's decision making. The developer is required to submit an environmental statement. The regulatory authority then considers the statement and representations made on it as a result of the consultation process. All the information amassed then comprises the environmental information that is used to make an environmental assessment of the impact of the project.

Within the planning process, the Directive's requirements have been implemented by means of the Town and Country Planning (Assessment of Environmental Effects) Regulations 1988 amended by the Town and Country Planning (Assessment of Environmental Effects) (Amendment) Regulations 1992. Other regulations were subsequently introduced in 1995 to deal with projects that fall within the scope of the Directive, which would otherwise be considered permitted development within the UK planning system. This makes it clear that there are no permitted development rights for the largest or most potentially damaging projects.

Projects that fall outside the planning system, such as afforestation, land drainage, salmon farming, highway schemes, harbours, electricity and pipe lines, have their own separate mechanisms for implementing the Directive introduced by separate sets of regulations. Space allows for only those projects dealt with within the planning system to be considered here.

What is regulated within the planning process?

The planning authority (including, in appropriate cases, such as call-ins and when dealing with appeals, the Secretary of State) must not grant planning permission in respect of an application to which the environmental assessment provisions apply, without first taking environmental information into consideration. This requirement applies to all types of development that fall within schedule 1 to the 1988 regulations. The schedule sets out types of project but in many cases only requires assessments for projects over a certain size or scale. The types of project listed are: crude oil refineries, large thermal power stations, nuclear power stations, storage or disposal facilities for nuclear waste, iron and steel smelting plants, asbestos plants, some chemical installations, motorways and other large roads, long-distance railway lines and large aerodromes, trading ports, large inland waterways, and waste-disposal facilities for incineration and treatment or landfilling of special waste.

A project belonging to a category described in schedule 2 to the 1988 regulations would only require an environmental assessment to be undertaken if it 'would be likely to have a significant effect on the environment by virtue of factors such as its nature, size and location' (regulation 2 of the 1988 regulations). In schedule 2 the types of development are listed under the following headings: agriculture, extractive industry, energy industry, metal processing, glass making, chemical industry, food industry, textile, leather, wood and paper industries, rubber industry, infrastructure projects, and a range of 'other projects', modifications of schedule 1 developments and schedule 1 developments solely for development and testing of new methods or products. Further categories have been added to the schedule 2 list, including: water treatment plants, wind generators, motorway service areas, coastal protection works, golf courses and privately financed toll roads.

Assistance is given to local planning authorities in determining whether schedule 2 projects require assessments within Department of the Environment circular 15/88. It is suggested there that an environmental assessment will be required in three main types of case:

- for major projects that are of more than local importance;

- occasionally for projects on a smaller scale that are proposed for particularly sensitive or vulnerable locations;

- in a small number of cases, for projects with unusually complex and potentially adverse environmental effects, where expert and detailed analysis of those effects would be desirable and would be relevant to the issue of principle as to whether or not the development should be permitted (paragraph 20).

The guidance suggest that the number of schedule 2 projects requiring assessment will be small. The Secretary of State has power to direct that an environmental assessment be provided, either overruling a local planning authority decision not to require one or in a case where a local planning authority has not considered the matter.

It has been held by the courts that the question of whether the proposed development falls within the scope of the regulations is broadly one for the local planning authority. This is challengeable by way of judicial review, but a challenge will only succeed where it can be shown that the authority's decision is irrational, unless the authority was mistaken as to the nature of the application and it is in fact a mandatory schedule 1 project.

Preparation of environmental statement

Once it has been accepted that an environmental statement is required, public bodies must make information available to the developer, if requested, in order to assist with the preparation of the environmental statement.

The 1988 regulations provide that an environmental statement must include:

- a description of the development, comprising information about the site, design and size or scale of the development;

- the data necessary to identify and assess the main effects that the development is likely to have on the environment;

- a description of the likely significant effects, direct or indirect, on the environment of the development, explained by reference to its possible impact on human beings, flora, fauna, soil, water, air, climate, the landscape, the interaction between any of these, material assets, and the cultural heritage;

- where significant adverse effects are identified, a description of the measures envisaged to avoid, reduce or remedy these effects;

- a non-technical summary of the statement.

Amended planning application procedure

In order to clarify whether a project will require an environmental assessment, the regulations allow for a formal pre-application request to be made to the planning authority for an opinion as to whether an environmental assessment is required. The authority must respond within three weeks, if requiring an assessment giving full reasons. If the applicant is unhappy with the local planning authority's view, or one is not provided, the matter can be referred to the Secretary of State for a direction. Even if the authority indicates that no environmental statement will be required, if it subsequently becomes aware that the proposed development is nevertheless a schedule 1 application or falls within schedule 2 and is likely to have significant environmental effects, it may (in he case of a schedule 2 project) or must (in the case of a schedule 1 project) require an environmental statement.

On receipt of a relevant planning application, the local planning authority must determine whether an environmental impact statement is required as part of the application. If no environmental statement has been provided with the application, and the authority considers one is required, it must give its written reasons within three weeks. The developer can than either indicate its acceptance of the need for a statement or inform the authority of its intention to seek a direction from the Secretary of State. If neither of these courses is followed, after a further three weeks the planning permission sought is deemed to be refused with no right of appeal to the Secretary of State. Where the developer indicates he/she will provide a statement, the planning authority must wait until the statement and certificates confirming that it has been publicised, and statutory consultees notified, have been provided. Unless the authority intends to refuse permission, it must allow 21 days from the date of receipt of the statement before determining the application.

A copy of the environmental statement and a non-technical summary must be placed on the planning register by the authority, and copies must be made available at a reasonable charge. The planning authority has the power to seek further information to supplement the environmental statement or evidence to verify the statement. The statement plus responses from consultees comprises the environmental information that must be taken into consideration by the planning authority before permission can be granted. Local planning authorities have 16 weeks instead of the usual eight weeks to consider applications where an environmental assessment is required. This period commences from the time the environmental statement is received. The Secretary of State must be notified of the determination of any application involving an environmental assessment.

Failure to have regard to environmental information in a relevant case will leave any planning permission granted open to challenge in the High Court.

APPEALS

The various procedures that may be adopted for planning appeals are also used, with some modification, in a range of other statutory environmental appeals. These include such matters as integrated pollution control authorisations, waste-management licensing, highways matters and consents to discharge to controlled waters. Much of the advice in this section will therefore also be relevant to fighting those kinds of appeals.

In general, objectors who want to fight a planning appeal effectively should:

- develop a plan of action early on;

- collect as much information as possible;

- make sure the information is accurate;

- understand the case being made by the other parties;

- liaise with other objectors and divide up the elements of the case;

- focus on key issues – it is usually a mistake to try to cover all points superficially;

- generate local support for the case through the media;

- consider a legal challenge immediately the decision is produced.

What are planning appeals?

Applicants who are dissatisfied with the decisions made on their planning applications by local planning authorities can appeal to the relevant Secretary of State. The application is then considered afresh in the context of national and local planning policies and other local factors with the responsibilities of the Secretary of State being carried out, in his/her name, by civil servants. The appeals process is handled by the Planning Inspectorate, an agency of the Department of the Environment, Transport and the Regions, based in Bristol, or the Welsh Office Planning Department in Cardiff.

Only the person who applied for planning permission originally can appeal against the decision of the local authority; there is no right of appeal by objectors. Their only remedy is to seek a judicial review in appropriate cases. The first step for objectors is to find out about the planning appeal. Usually objectors would have been involved in the process of persuading the local authority to refuse permission, or impose conditions on a planning permission that are unacceptable to a developer. Objectors who have been involved in this process should be contacted by the local planning authority once there is an appeal, as will a number of people occupying property around the appeal site. It would be unwise to rely on this, however, and where objectors are particularly concerned about a site they should write to the local planning authority to ask that they be advised in the event of an appeal being made.

What form will the appeal take?
Evidence relating to an appeal can be gathered in one of three ways:

- written representations;

- informal hearing;

- public local inquiry.

In each case the evidence collected from the local planning authority, the developer, statutory and non-statutory consultees and any other objectors will be considered by an inspector. The principal parties – that is, the 'appellant', or person making the appeal, and the local authority – will have an opportunity to comment on all the evidence and other representations made. Following a site visit, in almost all cases the inspector will produce a decision letter with reasoned justification. In the few remaining cases a report is produced by the inspector, which is sent to the Secretary of State with recommendations, and the decision is then made by the Secretary of State. Part of the procedure for dealing with appeals is laid down in statutory rules but to a large extent the procedure is determined by the inspector appointed to deal with the appeal.

Over 80 per cent of appeals are dealt with by way of written representations. However, if either the appellant or the local authority requires to be heard, a public inquiry or a hearing must be held to deal with the appeal. The Department of the Environment, Transport and the Regions can itself decide to hold an inquiry. Objectors cannot demand a public inquiry but if a substantial number of objections are received and the matter receives a lot of coverage in the local press it is likely that a public inquiry, or at least a hearing, will be held. Objectors should write to the local authority and the planning inspectorate to press for an inquiry if it is considered appropriate. Objectors should be careful about pressing for a public inquiry when the opposition to a development is weak. It may just serve to highlight the lack of opposition to the proposal.

Public inquiries are a very effective way of dealing with complex issues in great detail. Any public objections can be fully considered by the inspector and the evidence on which the inspector makes a decision or recommendation can be tested by cross-examination. However, a public inquiry will often cost all parties substantial amounts of time and money, imposing a great strain on the resources of the objectors. Holding an inquiry will cause further delay in determining the outcome of the appeal. In some circumstances this may be an advantage to objectors – for example, in cases where the success of the project depends on its speedy completion.

On what criteria are appeals decided?
Appeals are decided on the same legal basis as the original planning application to the local authority. However, inspectors appointed on behalf of the Secretary of State will give particular weight to government guidance (see the section on planning applications, page 117 above).

What can objectors do?
Planning appeals can deal with issues ranging from very straightforward simple questions, such as the granting of permission for a rear extension to a house, to very complex large-scale development with potentially massive environmental and social impact. In all cases, however, it is essential for objectors to be clear about the planning basis on which the appeal should be refused. Failure to clarify this central matter at an early stage will guarantee a failure to put a clear case before the inspector. As with objecting to planning applications before the local planning authority (see page 117 above), it is essential, once it has been decided to object, to clarify exactly what the proposal involves. Concerns to be raised with the inspector must be based firmly on the application before him/her. Vague concerns about matters that are not actually part of the application will be unlikely to have any weight in deciding the outcome of the appeal.

The importance of timing
Although the timescale may seem long at first, time inevitably passes very quickly, particularly for objectors, who tend to have very limited resources, and it is therefore essential to produce a timetable for all stages of preparation of the appeal as soon as possible.

The first stage should be devoted to the collection of as much information as possible not only about the application but also about the nature of the site, the potential impact of the process or activity proposed and the views taken by the local planning authority, statutory consultees and other bodies. In addition to the information available for planning applications, a particularly important starting point in cases where the local planning authority has determined the application will be the committee report on which the local planning authority based its decision.

It is essential to decide early in the process which issues should be focused on. In a complex appeal, there will be a number of important issues and it is usually beyond the means of objectors to deal adequately with them all. Those issues that will be properly dealt with by the local planning authority should rarely be duplicated, especially in relation to technical points such as traffic forecasts and housing requirements. Frequently, aspects such as pollution control and the impact on nature conservation will be less adequately dealt with by the local planning authority and in such circumstances it would make sense for objectors to focus on these issues.

Cases where the local planning authority appears unlikely to deal competently with any issues cause major difficulty. Also 'call-in' inquiries, where the local planning authority will not be leading the case for objectors, place other objectors in the position of taking a leading role. In these cases it may well be necessary to seek assistance from groups experienced in fighting planning appeals to cover all the aspects of the case required.

Preparing for a planning inquiry

The procedure adopted for the preparation for and conduct of the planning inquiry will be governed by one of two sets of rules. Which of these two sets depends on whether the appeal will be ultimately determined by the inspector. Since almost all appeals are delegated to an inspector, most attention will be paid to these types of appeals. The differences in appeals where the decision is made by the Secretary of State are dealt with later on in this chapter (see page 148).

Pre-inquiry timetable
The process starts with notification being given by the planning inspectorate to the local planning authority and appellant that a public inquiry will be held. The date on which this notification is given is called the 'relevant date', which is the beginning of the timetable. Within six weeks of that date the local planning

authority must serve a statement of case on the other parties. Not later than nine weeks after the relevant date the appellant must also serve a statement of case. Copies of these statements can be obtained by objectors from the local authority. They comprise a brief summary of the case that the principal parties will put at the inquiry with a list of documents on which they intend to rely.

In cases where it is anticipated that objectors will play a particularly large role, the inspector can require them to submit a statement of case also. In this event objectors will be given four weeks in which to produce such a statement.

In large or complex appeals where issues of procedure and the areas to be covered at the inquiry need to be clarified, the inspector may arrange a pre-inquiry meeting and/or provide a list of issues that are considered by him/her to be central to the inquiry. Objectors who propose to play a part in the planning inquiry should attempt to attend such a meeting. It will not only help to clarify the issues and procedure at the inquiry but it will also give a feel for how the process works and will assist the objectors in judging how best to intervene. If there are procedural or other matters that objectors wish to raise at the meeting, it may help to include them in a letter to the inspector at the planning inspectorate beforehand. It is worth noting that objectors are not able to have direct contact with the inspector at the planning inquiry, but other officials will be happy to discuss matters, which will be passed on to the inspector.

The local planning authority is required to publicise the inquiry. In addition to sending letters to objectors to the original planning application, a notice will be placed in the local press. However, objectors may feel that further publicity for the inquiry will be useful. This can be done by way of a press release and/or leaflets. Although objectors do not have an automatic right to appear at the inquiry, provided it is clear that they wish to make a positive contribution to the proceedings they will nearly always be allowed to present their case.

Every witness who intends to give evidence to the public inquiry is required to produce a proof of evidence. This is a written statement of the evidence to be presented. It must be given to the other parties and the planning inspectorate not less than three weeks before the beginning of the inquiry (unless other timetable arrangements have been made by the inspector). Since all parties must exchange proofs of evidence at this point, that will allow objectors a period of three weeks to prepare the presentation of their case. In practice these time limits tend to be flexible.

Preparation of evidence

Although the proof of evidence is only required for a public inquiry, a written statement will be needed, whichever form the appeal takes. It will generally be the same sort of document. In cases involving straightforward written representations these may just take the form of a letter, but otherwise it would be useful to adopt the following advice in all cases. The major document of every case presented at appeal is the proof of evidence. It is essential, therefore, that it clearly states the basis on which the objector considers the appeal should be turned down. A well-prepared proof of evidence will provide a good framework for the whole of the presentation of an objector's case.

A model proof of evidence, which clearly needs to be adapted for the individual circumstances of every case, is shown in the following table.

The witness This should describe the witness's technical qualifications for giving evidence. It should state, for example, that the witness is a local resident, mentioning the number of years he/she has resided in the area, and should give information on relevant involvement in other local activities, and relevant work or hobbies. If the witness is representing an organisation, then details of its aims, membership, etc., should be given here.

The site and its environs Each party will cover this area. It is surprising how frequently the description by local objectors differs from the other parties. It is important to focus particularly on the features that relate to the basis of the objectors' grounds for objecting to the development. For example, if a major aspect of the concern about the grant of permission for an industrial plant is the air pollution impact on local communities, it will be important to describe in some way the nature and location of the communities affected, making use of maps and photographs where appropriate.

The proposal   The proposed development should be described in a way that highlights the particular aspects of concern.

The planning policy context   This section should include relevant development plan policies, PPGs (government planning guidance) and circulars, supplementary planning guidance, etc.

The planning history   Description should be given here of any part of the planning history relevant to the objection.

Main planning considerations   This is the main part of the proof. Detail should be given of the basis on which the appeal should be turned down. It should be carefully referenced and only include propositions that can be substantiated.

Conclusions   There should be a clear summary of the main points of the case.

Appendices   Extracts of any documents referred to which the other parties are unlikely to have should be appended.

Where resources permit, it may be appropriate to have the evidence on different aspects of the case presented by different witnesses. Each witness must prepare a separate proof of evidence. It is important to ensure that proofs do not overlap or contradict each other. Where a proof of evidence consists of more than 1,500 words, a summary of not more than this length must be provided with the proof (although the summary is sometimes not produced until the day of the inquiry).

Presentation of the objectors' case

In general, objectors' cases are more effectively presented by committed lay people than by lawyers or technical experts. Provided the
local planning authority and/or other bodies are presenting the main
technical areas of the case, objectors themselves most effectively
present the case based on local knowledge and experience. Clearly, if
the objectors' case includes substantial areas of technical evidence
on complex matters it may be appropriate to employ professional
help to complement the campaigners' evidence. It is important to
remember that, even with cases dealt with at a public inquiry or
hearing, written representations can be made and submitted to the
inspector. If objectors have a witness who has something important
to say but who cannot attend the inquiry, it may be appropriate to
submit written evidence instead. It will, however, be given less
weight if it is not subject to cross-examination.

As has been mentioned above, matters that are not strictly
material to planning considerations will not be taken into
account by the inspector. However, a range of such issues may
help to create a context in which the inspector considers the
other evidence. Although a legal or technical expert may find it
difficult to present such evidence, it may be felt that a lay
witness on behalf of an objectors' group can effectively present
this broader context.

If, however, objectors intend to use a lawyer and/or expert it
is important to make the right choice. This is not always a
straightforward matter. Objectors should always question experts
carefully to establish that they have direct relevant experience
on the issues being dealt with and that their presentation will
be made in a way that will properly represent the objectors'
case. It is important that objectors clarify beforehand the area of
work that the expert and/or lawyer intends to cover and the cost
involved. If objectors have limited resources but nevertheless
feel that professional involvement is important, they should not
hesitate to seek to find sympathetic professionals prepared to
work for a reduced fee. To some extent, the need to raise funds
to pay for experts and/or lawyers can play an important part in
building a local campaign and publicising the objectors' case.

At the inquiry

The purpose of the inquiry is to test the evidence set out in the proofs of evidence, not to produce new evidence.

The form of the inquiry usually follows a set pattern. The inspector will open the inquiry at the time set out in the notice. It will usually be held in a room or council chamber at the town hall or in a village hall. It is usually laid out with a top table for the inspector and one table on each side of him/her for the representatives of the local planning authority and the appellants. There is usually a separate table where the witness giving evidence sits. Objectors will need to ensure that arrangements have been made for a place for them. Otherwise they will sit with members of the public, usually in rows of seats slightly apart from the parties giving evidence.

The inspector will open the inquiry by introducing him/herself and will describe the appeal. He/she will make a list of those members of the public who wish to speak, and confirm with the local authority that a site notice has been displayed and that notifications of the hearing have been sent to all objectors.

The appellant's advocate will stand up and open the case, usually quite briefly.

The first of the appellant's witnesses will be called and will go to the witness table. Usually the summary proof is read out, although additional points are frequently added as the witness reads through the proof, at the prompting of the appellant's advocate. This process is known as 'examination-in-chief'.

Other parties will then have an opportunity to ask questions of the witness, known as 'cross-examination'. The aim of cross-examination is to challenge the evidence put forward by a witness and to bring out other points that may be of use to the cross-examiner. Questions can be asked on the whole proof, not just the summary. The local authority advocate will usually ask questions first. If objectors have any questions they wish to ask they should indicate that they wish to do so at the end of the local authority cross-examination.

The inspector may then ask some questions of the witness and then, finally, the appellant's advocate will have an

146 ENVIRONMENTAL ACTION

opportunity to ask the witness to deal with any of the matters raised in cross-examination, known as 're-examination'. New matters cannot be raised at this stage because the other parties would otherwise not have an opportunity to ask questions on the new issues. Each of the appellant's witnesses is dealt with in this way. The local authority advocate does not make an opening speech but calls his/her witnesses in the same fashion as the appellant.

After the principal parties have given their evidence, others will be allowed to present their cases. These can range from an individual making a statement to a further full case presented by another advocate. Any witnesses making a statement or reading a proof may be subject to cross-examination, and so on as above. If, as is usually the case, a single person presents the evidence for a third party, without an advocate, the process is somewhat modified. The evidence-in-chief takes the form of a statement. Cross-examination is as above and re-examination merely allows an opportunity for the objector to make clear any points that he/she might feel was unclear at the end of cross-examination. The order of witnesses is not necessarily fixed. If, for example, an objector or an objector's witness can only appear on a certain day, the inspector will usually accede to a request for that person to give his/her evidence on that day.

An important point for lay advocates to remember is that it is not permitted for 'leading questions' to be asked of witnesses in examination-in-chief or re-examination. This means that advocates are not allowed to ask questions that indicate to the witness the answer that is sought. An example of a leading question is: 'Do you agree that the development is likely to harm the habitat of the local great crested newt population?' However, leading questions form an essential part of cross-examination.

Each party is given a chance at the end of the evidence to make a closing statement. The local authority usually goes first and the appellant always goes last.

The following are key points to remember about presentation at an inquiry:

• Detailed preparation is the essence of a good case.

- It can be useful to have one objector as the advocate and one giving evidence. The advocate is then in a position to stand back during cross-examination to see where clarification is needed.

- At lunch time or any other break in the proceedings a witness is not allowed to speak to his/her advocate, while he/she is giving evidence.

- It is important to remember that the inspector is the focus of the case. He/she will be producing the report. Advocates and witnesses should watch the inspector's response to the evidence and should be sensitive to any reaction, including allowing the inspector to complete his/her note. Lay advocates and witnesses may well feel awkward with the procedure but, provided they are making an effort to present the case in a polite and reasonable way, allowances will be made.

- Particular attention should be paid to answering questions put by the inspector.

- Objectors should not save up good points to make in response to cross-examination. Frequently, appellants' advocates will avoid any cross-examination of third parties wherever possible.

Advocacy is a particular skill that barristers spend years perfecting. Without that experience there are a few pointers that might assist the campaigning advocate. It is essential to be familiar with the case and have an idea of how it will be presented. Witnesses should be called in the logical order that makes sense of the overall case. Any supplementary questions to be asked of the witness during examination-in-chief should be agreed with the witness beforehand. Questions of a witness should not be asked if the inspector is still writing down the answer to the last question. Cross-examination should be very carefully planned, with a list of the questions to be asked and the order in which it is proposed to ask them being written down beforehand.

If objectors have questions to ask on issues not dealt with in any of the appellant's proofs, it should be clarified with the first relevant witness whether or not he/she will deal with that aspect of the case and if not who will. In asking questions in cross-examination it should be ensured that the answer is a full response and if not the question should be repeated. In the event of repeated failure to answer the question the inspector's assistance should be sought.

Closing remarks

Preparing a closing speech can be very difficult as its contents will arise out of the evidence given at the inquiry. A lay advocate should try to collect the points to be made as the inquiry goes along. A short adjournment of 10 or 15 minutes can be requested to allow preparation of the speech. There is no point in repeating all the evidence given by the witnesses. This is an opportunity to draw all the strands of evidence together to set out the campaigners' case as a coherent whole.

Conditions

The inspector will ask all parties to produce a list of conditions that they consider should be imposed on the permission if it is decided to grant permission. This request is frequently misunderstood by objectors to mean that the inspector has decided to grant permission or that if objectors prepare such a list of conditions they consider that, provided their conditions are imposed, the development would be acceptable. Neither of these is the case. However, if the inspector does grant permission, it may be important for the effect of the development to be limited by stringent conditions that objectors may be able to suggest.

Secretary of State decisions

Under 2 per cent of all appeals are not delegated to an inspector for a decision. The role of the inspector is to produce a report with recommendations to the Secretary of State. Another civil servant in the Department of the Environment, Transport and the Regions or Welsh Office will then prepare a decision letter and sometimes a minister's view will be sought as to whether it is acceptable. These cases tend to be the most important and politically sensitive ones. In these cases, pre-inquiry meetings are more likely to be held and the Secretary of State will give all parties a statement of the issues that are considered to be of particular importance to be examined in the inquiry along with notice of the meeting. The local authority and developer will be required to serve their statements of case four weeks after the conclusion of the pre-inquiry meeting. If there is no pre-inquiry meeting, both the developer and the local planning authority will be required to submit their statements of case six weeks after the relevant date. Not later than 12 weeks after the relevant date the Secretary of State will serve a written statement of matters to be considered at the inquiry.

Informal hearings

In certain cases the Secretary of State considers it best to hold an informal hearing. As the name suggests, there are no formal rules for informal hearings, which will be determined by the inspector. A code of practice produced by the Department of the Environment, Transport and the Regions suggests that such a hearing is inappropriate if:

• many members of the public are likely to be present;

• the appeal raises complicated matters of policy;

• substantial legal issues are raised; or

• formal cross-examination is likely to be necessary to test opposing cases.

If the Secretary of State favours a hearing, the planning inspectorate will offer the principal parties the option, and if they agree a hearing will be held. A copy of the code of practice produced by the Department will then be provided to all interested parties, along with notice of the hearing date. Not later than six weeks from the decision to have a hearing the principal parties will provide a written statement of case. Objectors should also produce a statement containing the same information that would be found in a proof of evidence for a public inquiry.

The hearing will take the form of a discussion led by the inspector, with those present introducing themselves and the inspector summarising what is understood to be each party's case. The inspector will then list the main areas and issues on which further discussion or clarification would assist. The appellant or a planning consultant on the appellant's behalf will usually start the discussion (lawyers rarely attend informal hearings) and objectors will be able to join in the discussion and raise questions for other parties. The appellant will have the last word.

Frequently the hearing will be adjourned to the development site where the discussions will continue. This clearly provides an opportunity for practical features of the difficulties raised by the development to be pointed out. If this does not happen it is likely that there will be an accompanied site visit at the end of the hearing. Frequently the inspector will agree to a request by the appellant to advise the parties by brief letter of the outcome of the appeal, to be followed in due course with a formal detailed decision letter. These letters will also be sent to any other person involved in the hearing.

Written representations appeals

Upon receipt by the Secretary of State of an appeal, the appellant and local authority will immediately be notified of the date of receipt, known as the 'starting date'. Within five days of the starting date the local authority must notify consultees and others who have made representations on the planning application that is the subject of the appeal. Within 14 days the planning authority must return a standard questionnaire about the case to the Secretary of State with the documents required, which detail how the local authority has dealt with the case and on what grounds the local authority oppose the appeal. Unless this comprises the full local authority case, they may send a written statement no

longer than 28 days after the starting date. Failure by the local
authority to notify an objector or a consultee to the original
planning application within the time limit may result in grounds
of legal challenge to the inspector's decision. In such a case an
objector should seek further time to submit his/her view.

Objectors frequently respond to 'written representations' appeals
by way of letter. However, in all but the simplest cases, it is suggested
that the outline set out for a proof of evidence be followed. The
appellant then has 17 days to reply to the local authority questionnaire or additional representations, and the final cut-off date for
comments is five days from the date of the appellant's response to the
local authority statement of case. A site inspection will then be organised by the inspector and the principal parties will be asked
whether they wish to attend. If a site cannot be properly viewed
without access to the site, an accompanied visit must take place, and
both the appellant and the local authority have to be represented. An
objector who wishes to attend may do so.

Other matters

Campaigning
Objectors to a development will frequently want to have a broader
focus to their involvement than merely participation in the appeal,
whichever form it takes. This is particularly the case for larger developments heard at inquiries, where the inquiry can be an important
hook for wider issues, but it can also apply in other cases.

Objectors should consider the following campaigning points:

• Stimulate early interest in the matter when the appeal is
  lodged. Let the local media know about their involvement.
  Contact local MPs, councillors and other local organisations,
  explaining their position and asking for support.

• Use petitions to help to build local interest.

• Where fund-raising is necessary to assist the presentation of a
  case, use it as a focus for publicity – for example, to attract local
  news coverage.

- Use other key dates, such as the pre-inquiry meeting and the date of the inquiry, as additional focuses for media coverage.

- Try to gather a large number of people at the inquiry, at least for the first day. Thereafter, as many people should attend as possible to show the strength of local feelings.

- Ensure that the media who have covered the story know where to contact objectors for further comment as the inquiry progresses.

Site visits
Site visits plays an important part in planning appeals. Frequently, at the beginning of the inquiry the inspector will announce that he/she has already visited the site unaccompanied to gain an understanding of the context. In addition, however, there will usually be an accompanied site visit. Objectors should prepare their involvement in the site visit carefully. It can play a particularly important part in an objector's case. Objectors should decide which local aspects would give the best focus to the case and arrange for the appropriate access to facilitate this. As this may involve obtaining the permission of the neighbours it is sensible to approach them well in advance. It is essential to ensure that they are aware that the visit may well involve a number of people.

Except in the case of informal hearings, where the hearing is frequently adjourned to the site, an accompanied site visit is not a continuation of the inquiry. New evidence must not be introduced at the site visit. However, it is often possible to increase the impact of evidence by pointing out to the inspector matters that have already been raised at the inquiry from particular vantage points around the site.

Costs in planning appeals
The general rule is that each party bears its own costs of preparing and presenting its case at a planning inquiry. Costs are only awarded against a party – that is, they have to pay the costs of another party – in cases of unreasonable behaviour. It is reasonably common for appellants to seek costs against the

planning authority when it is considered unreasonable that they had refused permission or that they have acted unreasonably in the conduct of the appeal. Costs will be awarded to or against third parties only in extraordinary circumstances, and provided third parties do not behave in a completely and intentionally unreasonable way they will not face this problem. Applications for costs will be made at the end of the inquiry. They may also be made in hearings or in cases dealt with by way of written representations.

The result

In appeals that are delegated to an inspector, a letter will be received by all parties addressed to the appellant. In the case of a Secretary of State decision, his/her decision letter, addressed to the appellant, will be sent to all parties with a copy of the inspector's report including recommendations from the inspector to the Secretary of State. Objectors should not be surprised if it takes a matter of months (and in some cases years) for a decision to be made.

On the receipt of the decision, objectors should study it carefully as a matter of urgency. It will have been written carefully to avoid generating successful legal challenges. It should include the inspector's findings on all the important issues. It will also contain a standard paragraph to the effect that the inspector has considered those matters mentioned and all other relevant considerations in reaching a decision. It will conclude by confirming that the appeal was either allowed or dismissed.

HIGH COURT CHALLENGES TO PLANNING DECISIONS

The Town and Country Planning Act 1990 provides for a statutory challenge to the High Court against a range of planning decisions. It is essential to note that, unlike appeals to the Secretary of State, which involve a complete rehearing of the case, these challenges must be based on a point of law only. This means that a complaint that the Secretary of State or inspector gave too much or too little weight to material considerations will not provide the basis for a successful challenge. In this and many other respects, such a legal action closely resembles judicial review, which will be covered in detail in Chapter 9. It is worth stressing here that these are very technical matters and it is essential to take legal advice before proceeding. Issues of legal costs must also be carefully considered, as an unsuccessful challenge is most likely to result in a costs order against an applicant running to thousands of pounds. Similar statutory High Court challenges exist from other decisions of the Secretary of State, notably on compulsory purchase and highway matters. Much of what is said here will be relevant in those circumstances too, although it is essential to identify the varying time limits for challenge in each case, as they are strict and often short (see below).

In cases where no statutory challenge exists, decisions are then reviewable by means of judicial review. For example, no statutory challenge lies against a planning authority decision to grant planning permission or against certain decisions of the Secretary of State, such as not to call in a planning application or not to order a fresh planning inquiry.

Grounds of challenge and time limits

The validity of a development plan or its alteration or replacement (section 287 of the Town and Country Planning Act 1990), or a decision made by or on behalf of the Secretary of State regarding a planning application (section 288 of the Town and Country Planning Act 1990), can only be made to the High Court on one of two grounds. The first is that it is not within the powers conferred under the Act and the second is that any relevant requirements have not been complied with. The first ground includes questions of irrationality and illegality as considered in Chapter 9 (page 190). The two grounds can be difficult to disentangle. It can be important, however, to do so because, in order for the second procedural ground to succeed, it is also necessary to show that the interests of the applicant have been substantially prejudiced by the failure to comply with the requirement.

Such challenges must be made within six weeks of the publication of the first notice of the approval, adoption or replacement of a development plan, or within six weeks of the date of a decision letter for a planning appeal. It must be emphasised that these time limits are absolute. Thereafter plans or decisions cannot be challenged.

Challenges can be brought by any 'person aggrieved' by the decision or adoption of a development plan. Although this requirement used to be narrowly interpreted by the courts, more recently it has been held to include any person who would be entitled to appear at a planning inquiry. This clearly includes the local authority and the developer but also in most circumstances now includes objectors.

With regard to enforcement matters, similar rights of appeal exist under section 289 of the Town and Country Planning Act 1990. There are, however, important differences. A critical difference is that an application for leave must be made within 28 days of receipt of the decision letter – in this case the High Court does, however, have power to extend the time limit in exceptional cases. In enforcement cases, the right of appeal is limited to the appellant, the local authority and any other person having an interest in the land. There is no statutory right of appeal for 'an aggrieved person' as described above, and therefore any challenge by campaigners would be made by way of judicial review. The grounds of appeal under section 289 are broadly similar to those referred to above.

Unlike with challenges relating to planning application appeals and development plans, leave must be obtained from the High Court before an application in respect of an enforcement appeal can proceed. The leave requirement is very similar to that for judicial review. In general terms, the effect of a statutory challenge to an enforcement notice will be that the notice will be suspended pending determination by the High Court. The local planning authority does, however, have powers to seek an interim order from the Court that the enforcement notice or part of it shall have immediate effect.

Procedure

The action is commenced by way of a notice of motion, which is served on the other parties such as the Secretary of State, the local planning authority and the developer. Evidence, by way of affidavit (sworn statement), must be filed and served on the other parties in support of the application. Other parties can also serve affidavits and frequently do. The Secretary of State often does not file any evidence. The applicant's affidavit will obviously exhibit the decision letter or notice that is the subject of the challenge.

In enforcement cases where leave to apply to the Court is required, a short leave hearing will be held, as with a judicial review. Once leave has been granted and in all other cases, the matter will then be fixed for a hearing. Planning matters tend to come to trial much more quickly than judicial reviews because planning QCs are appointed as deputy High Court Judges to hear these cases. As with judicial reviews, the evidence is all presented to the Court in affidavit form. The hearing therefore takes the same form as a judicial review.

The effect of an order to quash a decision or adoption is to remit the matter to the decision-taker. In planning application matters it would therefore be remitted to the Secretary of State to reconsider and in development plan matters to the local planning authority. In the light of the Court decision they will then determine what further steps are necessary – for example, whether to reopen any public inquiry or hearing or ask for further information from the parties. Any appeal against the decision of the High Court to the Court of Appeal must be made within four weeks of the date of the perfection of a Court's judgment or order.

CONCLUSION

The planning process is one that campaigners opposing a particular development will need to understand fully as it is one of the primary opportunities of stopping harm to the environment before it even starts. Further, it is one area of challenge where there is not the concern of having to pick up the costs of the developers, although the flipside of this is that there is unlikely to be a source of costs for the campaigners to pay for lawyers and experts other than through their own resources.

This is an area, therefore, where campaigners would be well advised to become fully conversant with the way the system operates. The planning stage is a time when they need to put their greatest efforts into opposing the development.

8   Road Proposals

The last decade has seen a significant rise in the level of opposition to the government's road-building programme. Gone are the days when schemes could be pushed through quickly by the then Department of Transport, with expected public support. Although the roads programme has been drastically cut over recent years, virtually any new proposal results in widespread opposition.

Following a shifting tide of public opinion the Conservative government was forced to reconsider its pro-road-building policies and the once-mighty Department of Transport began a slow U-turn away from road-building and made tentative steps towards a more sustainable transport policy.

Following their election in 1997, the Labour government agreed to carry out a comprehensive review of the entire roads programme. All those schemes for which work had not commenced on their coming into office were made subject to review with a full reconsideration of the issues. The results of the review were promised by spring 1998. The government agreed to produce a white paper in the autumn of 1998 looking strategically at the whole transport issue, taking into account alternatives to road-building.

The legal framework that governs the public inquiry, and other stages through which a road proposal passes, is complex and the procedures are not always easy to follow. This chapter will explain the law and the procedures used in determining new road proposals. It will focus on how to use them to the objector's advantage and also give useful information for campaigners opposing particular new road proposals.

WHO BUILDS ROADS?

The Department of Environment, Transport and the Regions

The election of the Labour government in 1997 brought about a change in the structure of the government departments responsible for building roads. The Department of Transport and the Department for the Environment were merged to form the Department of Environment, Transport and the Regions, which took on the functions of both previous departments and some additional ones. The political head of this newly formed Department is the Secretary of State for the Environment, Transport and the Regions (referred to throughout this chapter simply as the Secretary of State).

The Department has responsibility for overall transport policy and for building motorways and trunk roads. It has regional offices which it shares jointly with the Department of Trade and Industry. These offices are responsible for individual trunk roads and motorway proposals, with the London office addressing policy making and criteria setting. The Department also approves grants to local authorities for certain categories of road building and transport investment and may use local highway authorities as agents for its own schemes.

The physical building of roads is generally carried out by private civil engineering contractors and consultants employed by the Department through a tendering process.

Local authorities

Local authorities are responsible for all other roads; they are the promoting authority for the majority of road proposals. The routes of local authority road proposals appear in development plans produced by county, district and metropolitan councils, and are detailed in the Transport Policies and Programme, a document produced annually.

Private developers

The last Conservative government introduced a scheme called the 'Design, Build, Finance and Operate scheme', which means that the private sector finances the design, building and operation of the road. This is discussed later in the chapter.

STAGES THROUGH WHICH A ROAD PROPOSAL PASSES

In the case of trunk roads or motorways, the Secretary of State is both the proposer of the road and the decision-maker. He/she is therefore the judge of his/her own cause. Civil servants of the Department give evidence at a public inquiry, on behalf of the Secretary of State, to an inspector nominated by the Lord Chancellor. The inspector then reports to the same Secretary of State, whose private office sends the report to the division of the Department that promoted the road and briefed the civil servants who gave evidence. That division then drafts a decision letter in light of the inspector's report.

Similarly, local authorities have the power to propose, give planning consent for and build new local non-trunk roads.

Trunk road schemes

Early stages

The first indications that a particular new road is being proposed can come from a variety of sources, including the Roads Policy White Papers, which are the main advance notice of Department schemes in preparation (available from The Stationery Office, see list of contacts, page 414). Statements from ministers or MPs about 'studies' or 'need for action' are also a warning sign. Local authority staff and councillors are a useful source of information, as are the general and environmental press.

The Department's regional offices hold Regional Annual Consultative meetings with local authorities to determine the routes of proposed roads. These are closed meetings but some councillors may be prepared to discuss their contents. Consulting engineers then prepare the 'road study', which examines the feasibility of a possible road scheme. Following publication of this study regional offices compile 'regional transport briefs', which are documents detailing roads proposals in a particular region.

Following this there is a period of public consultation which usually consists of an exhibition of alternative route plans arranged by the Department, a brochure of routes, a questionnaire and a period of time for comment.

Following consultation, the Department makes a decision about its preferred route and the Secretary of State is notified of this decision. He/she announces the route and issues press statements and circulars, which are sent to everybody who filled in a questionnaire during the consultation period.

Statutory publication
The law controlling statutory publication and public inquiries for roads promoted by the Department is governed by the Highways Act 1980. The first stage is for the Secretary of State to publish and invite objections to documents called draft orders. The most important of these is known as the 'line order' because it defines the centre line of the new road. The line order is accompanied by a side roads order (SRO), which covers the diversion of existing roads and paths and various other details.

Another order, the compulsory purchase order (CPO), empowers the compulsory purchase of land and is normally published after the centre line of the proposed road has been fixed, but it is sometimes included in the same package of orders. The Compulsory Purchase Order Rules 1994 govern the procedure for the publication of CPOs. If the CPO is objected to by the affected landowners, the Department may hold a public inquiry to determine whether it should be granted, notwithstanding the objections.

The Department may publish other orders, such as an order for consent to affect a listed building or conservation area, or to revoke an order that was made in the past but where the proposed road was not built. Most new trunk roads or motorways are divided into several sections and the Department will publish an order of each type for each section.

The Highways Act 1980 requires the draft orders to be published in local newspapers along with a notice of the right to object and the closing date for objections. The orders are only set out in the legal section, but the news pages usually pick up on any controversial schemes.

Everyone has the right to object to such an order. The Highways Act 1980 requires that grounds for the objection are supplied, but these do not have to be presented in any great detail. If local people want to object, they should start by writing a letter to the promoting authority stating that they are objecting and giving the grounds. The public notices specify the objection period, which is a minimum of six weeks, with the exception of it being only three weeks for CPOs.

The inquiry

Once the objection period is complete, the Secretary of State has a duty to consider the written objections and then to decide whether or not to hold a public inquiry. This is usually dependent on the number of objections received.

The inquiry must be announced at least 42 days before it starts, with the announcement giving the place, date and time of commencement of the inquiry and the name of the officiating inspector. The notice states that the inspector will 'hear representations by statutory objectors and, at his/her discretion, by any other person who desires to be heard'.

'Statutory objector' is defined in the Highways (Inquiries Procedure) Rules 1994 and, broadly speaking, consists of the local authority where the road is proposed and owners of land directly affected by the road. In practice, however, inspectors are usually prepared to hear virtually anybody who wishes to give evidence.

The Department sends objectors a copy of the booklet *Public Inquiries into Road Proposals: What You Need to Know*, published in 1995. This is essential reading for anyone intending to appear at the inquiry but it should only be treated as background information as it covers only superficially certain important areas such as cross-examination, dealing with official witnesses and obtaining information from the Department.

The inspector may hold a pre-inquiry meeting, which is intended to deal with procedural issues such as agreeing a programme for hearing objectors at the inquiry and resolving any questions of fact before the inquiry.

The inquiry follows soon after this meeting, during which the inspector hears from civil servants from the Department, objectors and others who wish to be heard.

After the inquiry

The inspector writes a report on the inquiry, including recommendations on whether the orders should be made or how they should be modified, which is then sent on to the Secretary of State, who considers the report and the objections. A decision on the proposals is then announced and the inspector's report is published. The Secretary of State then makes or does not make the order.

Local authority schemes

Pre-inquiry

Local authorities indicate future proposals in planning documents, committee papers and minutes, and these are often publicised by the press. Ministerial statements may also give hints about local road proposals.

In contrast to Department schemes, objectors can lobby county and district councillors and officers, who are more susceptible to public opinion than the Department. Committee reports indicate the existence of new projects long before any surveys or design projects are carried out, giving objectors more time to organise themselves.

The detail of local authority road schemes should be contained in the council's development plan (structure or local plans, or unitary development plan (UDP); see Chapter 7). Structure plans and the UDP equivalent should specify the network of major roads of more than local importance, set out any major improvements to the network and contain policies on priorities for minor improvements.

They are the main vehicle for examining and questioning the principle of strategic local road schemes. Local plans and the UDP equivalent should identify the detailed route of proposed roads and improvements and indicate proposals for new roads and other improvements of a non-strategic nature.

The local authority allows for consultation on the development plan, giving objectors an opportunity to challenge the road scheme before a public inquiry. It is a good idea to question the basis of public transport strategies at the structure plan examination in public, the detailed alignment at the local plan inquiry, and finally the need for and alignment of the scheme at the local public inquiry.

The local authority also produce an annual document called Transport Policies and Programmes (TPP). The purpose of this document is to implement the transport policies in the structure plan. It lists all future schemes and often discusses those that are more advanced in detail. In effect it represents a bid for cash (Transport Supplementary Grant) from the Department. There is no statutory public consultation for TPPs, but the Department's circular on TPP submissions for 1995–6 states: 'it would reassure the Department to know that an authority's TPP as a whole has the support of the majority of local people'.

TPPs undergo an annual cycle of production. Each spring the Department issues guidelines to local authorities on production of TPPs and by 31 July each authority is required to submit a TPP bid to the local government office. Objectors should write to the local office at this time, setting out any objections to the TPP bid. Lobbying officers and councillors on the TPP contents for the following year should ideally start in the autumn. Objectors should demand to be consulted on the draft version.

A feasibility study will be carried out into the new road and results are usually published in committee reports, which can be obtained from the authority or via a councillor. Sometimes briefs or instructions are altered during the to-and-fro of consultant–client discussion, so it is worth asking what changes have been made.

Local authority proposals have to be publicised, although the method is often left at the local authority's discretion. For the smallest proposals a site notice or neighbour notification is required. If the proposal requires an environmental statement, represents a departure from the development plan, or affects a conservation area, listed building or public right of way, then the local authority must advertise it in a local newspaper and provide a site notice.

The local authority carries out public consultation over the new scheme, usually following the procedure for Department road schemes (see below, page 116).

A decision on a route will finally be made as result of council members' debate. The preferred route will be announced through press reports and committee minutes.

Statutory publication

For local authority roads there is no line order equivalent to that published for roads promoted by the Department. The authority makes an application for planning permission to build the road under Regulation 3 of the Town and Country Planning Act (General Regulations) 1992. Since the authority is the planning authority, it effectively decides whether to give itself planning permission to build the road. This much criticised system is called 'deemed consent'.

Local authorities are required to notify the Secretary of State for the Environment, Transport and the Regions – or, for roads in Wales, the Secretary of State for Wales – of local road proposals that have not been subject to local plan procedures or that conflict with the development plan. The Secretary of State has the power to 'call in' the application in certain cases and determine him/herself whether planning permission should be granted. This is discussed in Chapter 7 (page 148).

Objectors to a particular proposal should, therefore, consider lobbying the Secretary of State to call in the application. To achieve this it is best to write to the Department at every stage to ensure the civil servants become familiar with the proposal and with the objections. Further it will ensure the name of the campaign group becomes known to the Department. If a public inquiry is held and it is thought desirable to put forward a non-road-building alternative, this request should be put in writing as early as possible, before the pre-inquiry meeting.

Local authorities make their own classified SROs and CPOs and these are submitted to the Secretary of State. This cannot happen until deemed consent has been granted, either by the local authority or by the Secretary of State, for a called-in application. There is a six-week period for objection against the orders; objections are made to the Secretary of State, who copies them to the promoting authority with a request for the authority to respond to objectors.

The promoting authority then formally requests an inquiry if it wishes the orders to be confirmed (a separate inquiry to that held for called-in planning applications). Following the inquiry the Secretary of State confirms or does not confirm the orders.

The inquiry

The inquiry procedure is similar to that for trunk roads, but the planning application or order is treated like any other form of development and the scheme is unlikely to be approved if the inspector recommends refusal of planning permission (if called in) or refusal of confirmation of orders.

PUBLIC CONSULTATION

As mentioned previously, when a scheme has been put into a programme there will be a period of public consultation.

Trunk roads and motorways

A Department consultation exercise usually consists of an exhibition showing alternative routes, brochures showing the routes, a questionnaire and a period for comments.

The plans are not detailed except in urban areas and the brochures are generally drafted in public relations language, often omitting important facts. The exhibition will normally run for at least two days, including an evening, in a local hall. The Department's engineers and administrative support staff will be present, along with the consulting engineers responsible for the detailed planning. Public meetings are rarely held.

The main problem with this consultation process is that it does not normally question whether there is a need for the road. Only the choice of route is up for discussion. More recently non-road alternatives have occasionally been included, but they tend not to have been thought through and are usually primarily 'do nothing' alternatives.

Another criticism of the process is that other routes or alternative solutions that are not at the exhibition are not given proper weight. If objectors want to put forward an alternative proposal it is important to publicise it as much as possible through, for example, leafleting, publicity stunts and meetings.

The results of the consultation process are often treated as a proper measure of public opinion whereas in reality they amount only to a straw poll, comprising a self-selected sample from households to which questionnaires were delivered and the people who attended the exhibition.

Once the consultation is over, the preferred route will be presented, in general terms, in a press statement. There may be significant changes when the draft orders are published.

Local authority schemes

The consultation for local authority schemes is similar to that for trunk road schemes, but there are some differences.

Local elected members have to take note of public opinion when giving their views in committee. Also the decision on the choice of route made after the completion of the consultation process results from officers' reports and public committee meetings, not just the consultation process itself. Many of the options emerge through the preparation of development plans and the annual TPP process so a lack of support for a scheme from councillors may prevent it going ahead. Lobbying them at this stage is therefore important.

THE PUBLIC INQUIRY

Statutory framework

For road schemes proposed by the Department, the procedure for public inquiries is governed by the Highway Inquiry Procedure Rules 1994. For local authority road schemes, where the authority has granted itself planning permission and the application is 'called in' by the Secretary of State, it is governed by the Town and Country Planning Act (General Regulations) 1992.

Where a local authority has made orders and the Secretary of State calls an inquiry to determine whether or not to confirm them, the procedure is again governed by the Highway Inquiry Procedure Rules 1994. An inquiry concerning confirmation of a CPO made by the local authority is governed by the Compulsory Purchase by Non-Ministerial Acquiring Authorities (Inquiries Procedure) Rules 1990. Where there are CPOs and other orders (such as SROs) the inquiry procedure is governed by both sets of rules together.

Putting the public inquiry into perspective

Of all the stages where a new road proposal can be defeated, the public inquiry is probably the least likely. The public inquiry is promoted by the Department as the forum for the public to have its say and take part in the decision-making process. However, in the late 1980s, the inquiries into 141 out of 146 road schemes approved the road proposal concerned and generally public inquiries have resulted in road proposals going ahead with only a few minor modifications suggested at the inquiry.

Campaigns that have succeeded in preventing road schemes from going ahead have generally not done so at the public inquiry stage. For example, the M1–M62 Link in Yorkshire, the Preston Southern and Western Bypass, and the Exeter Northern Bypass were all dropped long before the inquiry stage as a result of high-profile campaigning at an early stage. A well-known campaign against a proposed road through Oxleas Wood in South East London succeeded in halting the scheme two years after the second public inquiry – because as the process continued the campaign intensified, despite the unfavourable result of the inquiries.

Notwithstanding this point, public inquiries are still a very important part of a campaign. They are an excellent way of highlighting the objectors' case, providing it with a focus and giving an opportunity to attract media attention – and there is still the possibility of winning. Objectors have recently started to win the odd inquiry – for example, at Norwich where public transport alternatives were successfully put forward.

The following section should be read in conjunction with Chapter 7, which explains the procedures for local authority public inquiries (pages 136–48). This chapter concentrates on the points not covered in Chapter 7 that are of particular importance to road inquiries. More detailed information on public inquiries can be found in the comprehensive *Campaigner's Guide to Road Proposals*, available from the Council for the Protection or Rural England (see list of contacts, page 406).

Evidence relevant to road inquiries

Objectors can bring in witnesses such as experts and local people affected by the road. From the Department's point of view it is preferable if only one person from each group gives evidence because it makes it easier to discredit the group. However this may not be in the interest of the campaign; it may well be better for a few people in the group to give evidence, some concentrating on different technical areas and others giving more general, anecdotal evidence.

For Department schemes, the proposed road will be part of the national roads programme and therefore government policy. The object of the inquiry is not to challenge government policy by questioning the need for the road, but to measure the weight of local opposition. This means that campaigns need to consider carefully how the public transport alternatives are put forward and it is possible that the inspector may not allow the evidence to be given (see page 137–9). For local authority schemes, however, the procedure is governed by the Town and Country Planning Act 1990, and issues such as the need for the road and public transport alternatives can be discussed.

Solid concrete evidence of the environmental effects of the road and anecdotal evidence from people affected are the major points for the campaign to make. By far the most important arguments to push forward, however, are those challenging the government's justification for the road, including its form of cost–benefit analysis, COBA 9 (see page 174 below), since this is what the inspector will attach most weight to at the inquiry.

Pre-inquiry meeting

At the pre-inquiry meeting the scope of the inquiry is an important issue to be dealt with. The inspector may indicate an initial view of the scope of the inquiry and it may be important to try to extend this.

For schemes promoted by the Department, the inspector is likely to say that the inquiry cannot deal with three issues: government policy, law and compensation. This includes the question of whether there is a need for the road at all. A possible way round this is for the inspector to be asked to allow discussion of alternatives to the road – such as public transport, walking and cycling alternatives – on the basis that this is not challenging government policy because the government has produced policy encouraging this (for example PPG13 – see later in this chapter). It is important to question the need for the road as early on as possible.

Other than these restrictions, the inspector in all types of inquiries has very wide discretion as to what evidence will be heard and how much weight is attached to each piece of evidence.

The scheme may be part of a longer route than that being considered at the inquiry. The Department often divides long routes into a series of local 'bypasses' in order to take attention away from the overall impact of the route. If this is the case, it should be explained to the inspector that this scheme is part of something bigger and that the whole route should be taken into consideration by the inquiry. It might well be advisable for residents from other parts of the route to attend the inquiry to explain the effect it will have elsewhere.

It is usually possible, before the inspector, to argue that transcripts be made available of the evidence given at the hearing. The Department may resist on the grounds of cost, the response to which is likely to be that transcripts are essential because the objectors cannot be at the inquiry all the time whereas the Department can. They should be free and available the morning after the evidence is given.

If there are key objectors who cannot make daytime sessions, evening sessions should be requested. The inspector may offer them only for certain days but, if appropriate, it should be explained that certain objectors need to be present throughout the entire proceedings in order to follow, and respond intelligently to, the developing argument of the inquiry.

The Department is responsible for finding the inquiry venue and it may be in a place inconvenient to the objectors. If so, arguing for a venue that is easily accessible may succeed. The Department may argue that further local venues are not available but this can be challenged by independently checking on venues prior to the inquiry meeting.

Summary of procedures for inquiries

For local authority road schemes, where the Secretary of State has called in a planning application, the procedures are as described in Chapter 7.

For trunk road and motorways promoted by the Department, and local authority schemes where the authority requires confirmation of orders made by them, the procedure is slightly different and is set out below.

The barrister for the Department makes an opening speech and the Department's witnesses read their proofs of evidence, explaining any issues as required by their barrister. Other parties may then ask questions, including requests for further information, but may not cross-examine the Department at this stage. If there are any supporters of the scheme they then read their evidence, local authorities first of all, and are open to cross-examination by objectors.

Objectors then make opening speeches on their own cases. They may cross-examine the Department on its case before calling witnesses, although they will not normally be able to put their own case (for example for an alternative route) to the Department's witnesses at this stage. The Department may re-examine its witnesses.

Objectors either call witnesses or read their own proofs of evidence. Following this, objectors and their witnesses are cross-examined by the Department's barrister and any supporters of the scheme. The inspector is likely to ask questions from time to time to clarify matters as they arise.

Objectors are re-examined by their barrister if they have one. If not, re-examination is in theory not possible but they can ask to make a statement if it is only to clarify matters raised in cross-examination.

The Department then presents what is know as rebuttal evidence This is additional evidence dealing with what objectors have said, such as considering alternative routes. This is the first time the objectors will see this evidence and they should usually ask for a short adjournment so that it can be studied properly.

The objectors can cross-examine the Department's witnesses on rebuttal evidence and, if they did not take the opportunity to cross-examine earlier, on the whole of the Department's case. This is the objectors' greatest chance to discredit the Department. This is the point where the objectors' barrister (if any) should be used to challenge the Department on its evidence, particularly in relation to the cost–benefit figures (see below, page 174). If the Department does not have documentation to support them, an adjournment should be requested until the required documentation is produced.

Counter-objectors, that is people or organisations who object to alternative routes or anything else proposed by objectors, then read their evidence, cross-examine objectors if they did not do so earlier, are cross-examined by objectors, and are re-examined by their own advocate.

Counter-objectors, objectors and supporters then sum up their cases in 'closing statements', usually in the reverse order to that in which they gave evidence at the hearing. Finally the Department's barrister makes a closing speech.

Tactics at road inquiries

Before the inquiry, the Department's supporting documents should be obtained as early as possible. The Department may be reluctant to provide some documents – for example, early internal reports on the scheme called 'scheme assessment reports' – but the objectors should be persistent in their demands for such documents.

It is a good idea to gain the local authority's support for the objectors' case because they can provide useful assistance, including lawyers, traffic experts, environmental experts and photocopying. All documents used in the inquiry should be kept and any instances of procedural irregularities or unfairness should be recorded in case there is legal action at a later date.

It is a good idea to 'play by the rules' – for example, giving proofs of evidence in on time and being courteous and helpful to the inspector. The inspector should, however, be made aware of any irregularities by the Department, such as not producing relevant documents.

Roads inquiries often intimidate and confuse objectors. This can be counteracted by trying to keep them as informal as possible by, for example, bringing along children, hot drinks and anything that will make for a relaxed atmosphere. It is likely, however, to be counter-productive if objectors resort to disrupting the public inquiry.

If the Department makes unsubstantiated statements or if its figures do not add up, or the procedures are being used against the campaign, the media should be informed. The public inquiry is a media event and any negative publicity for the Department will strengthen the case. If the inspector is being very unreasonable this might be an issue also worth exposing.

Justifying a new road

There are four criteria that the promoting authority will use when arguing for a road scheme:

Traffic forecasts
These predict the level of future traffic on the road system in any future year. Currently, huge increases in traffic are forecast (between 83 per cent and 142 per cent between 1989 and 2025) and the forecasts are claimed to be 'policy' and therefore not challengeable at inquiries.

Traffic assignment
This allocates forecast traffic to different roads in a computer model. Assignment rarely includes generated traffic because the modelling methods are not equipped to deal with this.

Cost–benefit analysis (COBA)
This calculates the monetary value of time, operating cost and accidents saved by a new road.

Environmental appraisal
This is carried out using the Manual of Environmental Appraisal (MEA). Some attempts have been made to quantify measures of environmental impact, but in reality most decisions about environmental effects depend on the level of local objections.

Generally, the authority uses these four criteria to show that the proposed scheme will solve traffic problems and be good value for money.

KEY ISSUES FOR USE IN THE CAMPAIGN

There are a number of legal and other issues of which campaigners should be aware for use in arguing their case against the road at the public inquiry and when campaigning.

Cost–benefit analysis

The main economic justification for the roads programme, and individual schemes within it, is a form of cost–benefit analysis known as COBA 9.

When assessing options for road schemes, a study area is defined and the origins and destinations of traffic within that area are surveyed. A forecast of how fast traffic is expected to grow is applied so that predictions can be made of traffic flows for future years. The predictions are then applied to the road network both with the proposed new road and without it (the 'do minimum option').

COBA 9 chooses a hypothetical amount of money to represent the time saved by each vehicle that will use the new road. This sum is multiplied by the number of vehicles predicted to use the road to give a sum supposedly representing the total time savings made as a result of the new road. This figure is compared to the costs of road construction and maintenance and the hypothetical costs of road accidents.

If the sum representing the time saved by the new road is greater than the costs of the road and accident savings then the road is said to be justifiable under COBA 9. For most proposed road schemes, COBA 9 will 'prove' that the proposed road will show an economic return.

Weaknesses of COBA 9
The COBA 9 method has calculated that, on average, 80 per cent of the benefits of a road scheme are from hypothetical savings in journey time made by motorists avoiding traffic jams by using the new road. However, as the House of Commons Transport Select Committee pointed out, 'The assumption that small increments of time have real economic value when aggregated over a large number of vehicles is unsubstantiated.'

COBA 9 also assumes that vehicle occupants put the hypothetical saving of a few minutes journey time that building a new road supposedly brings before the destruction that road building causes. COBA 9 does not give any value to the areas of ancient woodland, listed buildings, and so on, destroyed to build the road.

The traffic forecasts are applied to the road network whether or not that road has the physical capacity to cope with the forecasts. In reality, however, traffic growth is constrained by the capacity of the road network. This approach allows the Department to compare the 'do something' scenario (a new road with smooth-running traffic) with the 'do nothing' situation (an impossibly congested network without a new road). Thus COBA 9 credits road proposals with the benefits of relieving traffic jams that could never exist.

COBA 9 does not take account of 'traffic generation'. Traffic generation refers to new journeys that are only made because of the existence of the new road. In the wake of the SACTRA report the Department of Transport has accepted that traffic generation occurs, and in future traffic modelling is supposed to take it into account.

The COBA 9 calculation produces either a financial profit or a loss. This is called the 'net present value' (NPV). NPVs are not assessed by the percentage return on investment but by size. Thus an NPV of £20 million representing a 15 per cent return is preferred to an NPV of £1 million representing a 45 per cent return, so large schemes are favoured.

COBA 9 is not applied to pubic transport proposals. Generally public transport schemes have to show a straight financial return before the Department of Transport allows operators to borrow money to realise a proposal.

Campaigning on COBA 9

COBA 9 is based on a series of hidden assumptions that may well not be justified. Road proposers use COBA 9 to 'prove' that their road is 'good value for money', and this is a powerful argument in convincing politicians and the public to support a road. The weakness of COBA 9 can be used by objectors to undermine the credibility of the arguments supporting the road they oppose.

In the instances where objectors have won at public inquiries, they have done so by challenging COBA 9 successfully and showing that the Department's calculations are flawed. COBA 9 is the heart of the public inquiry and objectors will ideally need a barrister and an expert who fully understands COBA 9 to challenge the calculations. The actual theory behind COBA 9 cannot be challenged at inquiry since it is government policy.

The SACTRA Report

In December 1994 the Standing Advisory Committee on Trunk Road Assessment (SACTRA) Report concluded that new roads generate new traffic. The implications of this for the roads programme are enormous.

Most new roads are justified on the basis that they will relieve congestion elsewhere on the road network. However, if new roads attract new traffic then that congestion relief is diminished or even cancelled out, which means that the claimed economic benefits from the road scheme are likely to be severely reduced.

Overall, SACTRA addressed four questions:

Is induced traffic a real phenomenon?
'Induced traffic can occur, probably quite extensively, though its size and significance is likely to vary widely in different circumstances.'

Does induced traffic matter?
'These studies demonstrate convincingly that the economic value of a scheme can be overestimated by the omission of even a small amount of induced traffic. We consider that this matter is of profound importance to the value for money of the road programme'.

When and where does induced traffic matter?
According to SACTRA, induced traffic occurs in three sets of circumstances: where the road network is already congested, where travellers are responsive to improvements in the road network and where a scheme causes big changes in travel costs. In practice this means traffic induction is most likely to occur on roads in and around urban areas, river crossings and capacity-enhancing inter-urban schemes, including motorway widening.

What needs to be done?
SACTRA states that there should be strategic economic and environmental area-wide assessment of road schemes. Corridor improvements (for example, south coast superhighway) should be assessed as a whole rather than divided into small sections as happens now. Also a more sophisticated form of traffic modelling (using 'variable demand methods') should be introduced that can take account of induced traffic.

The government accepted SACTRA's main conclusions that new

roads induce new traffic and issued internal guidance, the Guidance on Induced Traffic, to their road planners to take account of these findings. They have stated that COBA 9 still stands but that all roads must be looked at again in the light of SACTRA and that COBA 9 must take into account traffic generation. Variable demand models look likely to be phased in for schemes where induced traffic is likely to occur. Area-wide models can now be used to assess schemes that are part of a wider 'strategy'.

Objectors can challenge the Department's COBA 9 calculation in a public inquiry on this basis if traffic generation has not been taken into account.

Planning Policy Guidance 13 (PPG13)

Planning Policy Guidance Notes (PPGs) are documents that communicate government policy to local authorities. Because they are government policy, local planning authorities must take them into account when preparing development plans. The guidance may also be a material consideration for decisions on individual planning applications and appeals.

PPG13 provides advice on how local authorities should 'integrate transport and land-use planning'. The guidance states that local authorities should carry out transport programmes in ways that 're-duce growth in the length and number of motorised journeys, encourage alternative means of travel that have less environmental impact and hence reduce reliance on the private car'.

The guidance goes on to say that 'in this way, local authorities will help to meet commitments in the government's Sustainable Development Strategy to reduce the need to travel; influence the rate of traffic growth; and reduce the environmental impacts of transport overall'.

The guidance calls for more mixed-use development, with an emphasis on higher-density development planned along public transport corridors, an avoidance of major developments in locations not served by public transport, and encouragement of journeys by foot or bike.

It states that measures should promote choice in means of travel, reduce dependence on the car and increase attractiveness of

urban centres against peripheral development. There is stress on
the need to plan for cyclists and pedestrians, and of making rail
the focus for regeneration.

According to the guidance, new routes must, wherever possible,
be kept away from protected areas such as sites of special scientific
interest. Trunk roads must 'serve their purpose as corridors of
movement' and not 'have their strategic role undermined by
development which encourages their use for short local trips'.

Campaigning on PPG13

PPG13 demonstrates a real shift in government attitudes towards
a more sustainable approach to transport and it is a very powerful
tool that you can use in a public inquiry. Objectors should state
that the road proposals do not fit in well with the government
policy of reducing the reliance on the motor car.

Environmental impact assessment

Environmental impact assessment is a process that is meant to
provide decision-makers with an objective assessment of the
implications for the environment of approving major projects. In
theory it provides the opportunity to take environmental considerations into account in the decision-making process.

Under the European Council Directive 85/337/EEC, all motorways and expressways should be subject to an environmental
impact assessment (EIA). Article 3 of the Directive states that the
environmental impact assessment must 'identify, describe and
assess in an appropriate manner, in the light of each individual
case ... the direct and indirect effects of a project on the following
factors:

- human beings, fauna and flora

- soil, water, air, climate and the landscape

- the interaction between the factors mentioned in the first and
  second indents

- material assets and the cultural heritage.'

A non-technical summary has to be produced so that the public can have access to the information contained in the full environmental statement. The Directive also suggests that there should be extensive public consultation on the EIA (article 6, paragraph 3).

The procedure for EIA of Department schemes is set out in the Manual of Environmental Appraisal (MEA) along with Departmental Standard HD 18/88, and is as follows. When the Department decides whether a scheme should enter the roads programme it produces a 'Stage One EIA', which is a 'desk-top study of environmental constraints'. Prior to public consultation on route options the Department carries out a 'Stage Two Assessment' which is a 'broad assessment of environmental impacts and options'. It is not until the draft order stage that the Department produces the detailed EIA, which is a full environmental impact statement and non-technical summary conforming with the regulations.

Local authority schemes come under different regulations, but since there is little government guidance on carrying out EIAs other than the MEA and Departmental Standard HD 18/88, most local authorities follow these procedures.

Criticisms of UK EIA on roads
Department EIAs only consider carbon monoxide emissions when examining air pollution, ignoring carbon dioxide, the main greenhouse gas, of which transport is the fastest growing source.

Department EIAs are based on the difference between 'do nothing' (the situation without the proposed scheme) and 'do something' (the situation with the proposed scheme) scenarios. The Department forecasts that traffic will grow at the same rate under each scenario. Under this assumption the 'do nothing' scenario will be a road network clogged with traffic, leading to high levels of air pollution. But under the 'do something' scenario, traffic will flow freely, and air pollution will as a consequence be less severe than in the 'do something' scenario. Thus many road scheme EIAs claim that air pollution will decrease if the road is built!

Department EIAs do not consider alternative scenarios: for example, comparing the environmental consequences of the proposed scheme with improved public transport and the introduction of area-wide traffic calming.

The Secretary of State is required under the UK regulations to produce an environmental statement for the preferred route at the time when draft orders are published. However the Directive states that governments should 'affirm the need to take effects on the environment into account at the earliest possible state in all the technical planning and decision-making procedures ...'. Under Article 6, paragraph 2, member states shall ensure that 'the public concerned is given the opportunity to express an opinion before the project is initiated'. As soon as a road scheme is added to the British road programmes it has been 'initiated', as from then on it is government policy to build the road. Thus, by not publishing the EIA until the preferred route or draft order stage, the Department is failing to ensure that the public is given the opportunity to express an opinion before the project is initiated.

Where there are a number of schemes on a road corridor, each scheme is assessed independently. There is no strategic or cumulative EIA. For example, widening the M25 has enormous environmental implications for the Southeast, yet these impacts will not be evaluated. Instead the widening has been split up into 18 different projects, each of which gets an independent EIA.

Campaigning on EIAs

Exposing deficiencies of EIAs to local people, press and politicians will help to win support for the objectors and strengthen their case. They should also press the promoting authorities for an environmental statement to be produced.

A complaint about the deficiencies of an EIA can be made to the European Commission by writing a letter to the EC Environment Commissioner.

Shadow toll roads and DBFO

The Conservative government indicated that most, if not all, new major roads would be financed under a system of 'shadow tolling'. Under this system, a road scheme from the national roads programme is offered to the private sector as a Design, Build, Finance and Operate scheme (DBFO). This means that the private sector designs, builds, finances and operates the road and is recompensed through the payment of shadow tolls. For every vehicle that uses the road, the treasury pays the company an agreed amount.

The Labour government intends to continue the principle of private finance for road schemes, though the exact method of operation may be changed in the future.

Shadow tolling is a form of hire purchase. The government pays for the road over a period of time. How much the government pays in total depends on how much traffic uses the road and how much the government pays per vehicle. The uncertainties in traffic levels mean that there is a risk involved for both parties, and evidence suggests that to ensure private sector involvement it is the government, rather than the private sector, that will take on most of the risk. In other words, shadow toll roads will cost the taxpayer more in the long run than conventional contracts.

It will be in the interests of DBFO scheme constructors to encourage as much traffic as possible to use their roads. The more vehicles (and by extension the more air pollution, polluted run-off, noise, etc.) the more taxpayers' money they receive. This inducement to traffic growth runs contrary to the recommendations of the Royal Commission on Environmental Pollution and the emerging direction of government policy.

In addition, the DBFO schemes will add to pressure for the development of greenfield sites next to the new roads, contrary to Department planning guidance.

Campaigning on DBFOs

If a DBFO is used, the COBA 9 calculation will be affected because the road is financed in a different way. Objectors (or an expert) should question the Department's estimates of the number of cars predicted to use the road in its COBA 9 calculation. COBA 9 is not sophisticated enough to take the factors introduced by DBFO schemes into account.

Public transport alternatives

When presenting evidence of viable public transport alternatives to Department schemes, objectors should stress that it is accepted government policy to encourage walking, cycling and public transport use. Objectors should clearly state that they are not challenging the need for the road as such, simply offering better alternatives.

It is advisable to prepare outline alternatives only – if a detailed alternative is prepared the objectors are laid open to challenge by the Department or local authority on the details of the proposal.

General campaigning tactics

Any campaigning should commence as early as possible – that is, as soon as details of the road scheme are known. It is appropriate for campaigners to try to win the battle before the public inquiry starts. For a full discussion of direct action and other campaigning tactics, see *Road Raging: Top Tips for Wrecking Roadbuilding*, available from Road Alert! (see list of contacts, page 410).

The strategy is ideally to establish a good relationship with the local media and to come up with a story at quiet times, such as Christmas, Easter and Bank Holidays, or put on eye-catching stunts, which are often easier to organise than mass rallies. Campaigners should cultivate local politicians from all parties; inviting them to join the campaign can help.

Putting forward alternatives to the road scheme based on public transport or cycling can be important. Having such alternatives shows that the objectors are not only opposing the road scheme but also have positive solutions. The Department or local authority is then under pressure to come up with reasons why the scheme is unacceptable.

Campaigning should be on a regional and national basis. The Department has often attempted to promote road schemes as being locally specific, despite the fact that they form part of a wider picture. While still retaining the local identity of the inquiry, it is often helpful to form regional or corridor alliances for coordinating regional stunts, cultivating regional press and bringing together MPs from a wider area.

The most successful pressure group campaigning tends to be pro-active, setting the agenda, all the time fighting on its own territory.

CONCLUSION

The widespread public protest that has grown in the last ten years, particularly the direct action movement, forced the government to reconsider its unsustainable position of attempting to build its way out of the transport problems facing this country. In confronting these problems, society has to deal with wider questions about what we value in our lives. The anti-road movement has questioned fundamental assumptions, such as the link between road-building and economic growth, and the relief of congestion by building new roads. Campaigners have consistently won the arguments. It is now clear that what we need are sustainable, long-term solutions that improve the quality of life for all members of society and that safeguard the planet as a whole.

This chapter has outlined the law and procedures set up to determine road proposals, though they are considered by many to be biased and unjust. To campaign effectively it is important to 'know your enemy', and this chapter is intended to provide campaigners with a thorough understanding of how the system operates.

This is an exciting time for the anti-road movement, and a new government has brought fresh opportunities. It has the chance to continue the move away from a selfish, car-owning society that plans only for the short term, by making further steps towards the development of a more caring, community-based, sustainable transport policy. This would mean putting the quality of the environment in the widest sense before the need for an individual to get from A to B, cocooned from the outside world in the false security of the motor car. By using the information in this chapter, and the contacts at the end of this book, the campaigner can, with a little determination and ingenuity, bring this vision closer.

9   Judicial Review

WHAT IS JUDICIAL REVIEW?

Judicial review is a significant weapon in the armoury of environmental campaigners, albeit one that is only usually turned to as a last resort. It can be a great help in attempting to prevent action that will harm the environment because, unlike most of the other legal tools in this book, judicial review is usually aimed at stopping decisions before they are put into effect rather than when it is too late and the harm has already been caused.

That isn't to say that a judicial review is easy or guaranteed to succeed. It certainly isn't short of many other problems, not least being the funding of the action. There are ways round at least some of these problems, however, and any campaign group should always consider the possibility of such a review where a decision to permit a development to go ahead has been made.

Judicial review allows the courts to review the lawfulness of decisions made by public bodies – including government departments such as the Department of the Environment, Transport and the Regions, regulators such as the Environment Agency, and local authorities carrying out their various functions such as town and country planning and pollution control. This type of action has grown dramatically as a safeguard over the past two decades. The judiciary has also been expanding its influence in this area, largely because of its concern about the unfettered way in which the government exercised its powers during the lengthy period of one-party government and large parliamentary majorities. For the same reasons, campaigners who have been traditionally more accustomed to expending their time and effort on lobbying and campaigning to change the government's view have found that their concerns have fallen on deaf ears. As a result, they have felt they have no choice but to resort to the courts for assistance.

It must be said from the start that taking a judicial review action is no miracle cure. Apart from in certain European Union matters, Parliament makes the laws and it is the courts' job to ensure that they are carried out. If the government can muster a majority in Parliament and is determined to have its way, the courts will not be able to intervene. Equally, if a local authority makes a decision that is consistent with the law and government policy, albeit not a decision the courts would favour, they cannot interfere.

Furthermore, even if a judicial review is successful, the courts will not replace the decision complained about with a decision the campaigners prefer. The court will merely order the decision-taker to reconsider the decision in a lawful way. Since frequently the flaws in the decision are procedural ones – for example, that proper consultation was not carried out – provided the public body carries out the procedure correctly it is quite possible that the same decision will be repeated.

In many cases, however, campaigners believe that, although ultimately the decision-taker could come to the same decision, the surrounding circumstances makes this unlikely. The process of legal challenge is likely to attract a great deal of bad publicity for the public body, which may well bring many other objectors together and provide a large tide of opinion that the public body feels unable to ignore. Further, quite often proposals are very time-sensitive and in such cases, if the decision is delayed for the significant period it takes for the case to go to court, the offending activity may no longer be a realistic option.

A further factor that can be a problem about judicial review is that often the environmentally harmful activity is not being carried out by the public body against whom the judicial review is taken. The complaint is that the public body or regulator either allows that harmful or polluting activity to be carried out or does not prevent it from continuing. This can move the focus of the campaign away from the polluter onto the regulator, which may not always be a helpful part of the campaign. On the other hand, once a regulator recognises that there are effective campaigners scrutinising its activities, it may well decide that it is necessary to be more rigorous in carrying out its enforcement obligations over polluters.

This chapter summarises the circumstances in which the taking of a judicial review is appropriate and how it can help to protect the environment. It will then outline the process by which a judicial review can be brought and the stages involved. This is undeniably a complex and specialised area of the law. Although it has been undertaken successfully by individuals or groups on their own behalf (known as litigants in person), most environmental campaigners would recognise that it is an area requiring not just legal input but input from specialist barristers and solicitors with real experience of this type of case.

The parties to a judicial review are referred to in special terms. The body or person bringing the judicial review is known as the applicant. The body whose decision is under challenge is known as the respondent. The title of a case where, for example, Friends of the Earth is judicially reviewing a decision by the Environment Agency would be: *R.* v. *The Environment Agency*, ex parte *Friends of the Earth Ltd.*

The following are examples of successful environmental judicial reviews.

- *R.* v. *Secretary of State for the Environment*, ex parte *Royal Society for the Protection of Birds* (1997). Planning permission having being granted to develop Lappel Bank, the Secretary of State decided to exclude it from the Medway Estuary Special Protection Area designated under the Wild Birds Directive 1979 on the grounds that economic considerations outweighed environmental ones. The European Court of Justice held that the Secretary of State was not entitled to take account of economic considerations when designating a special protection area under the Directive.

- *R.* v. *Secretary of State for the Environment*, ex parte *Kingston upon Hull City Council* (1996). In designating the Humber and Severn Bridges as the estuary boundaries under the Urban

Waste Water Directive (and thus requiring a lower standard of treatment of sewage effluent discharged), the Secretary of State had wrongly taken account of the cost of sewage treatment and as a result avoided the obligations imposed by the Directive.

- *R.* v. *Carrick District Council,* ex parte *Shelley* (1996). The Council's failure to serve an abatement notice under section 79 of the Environmental Protection Act 1990 in relation to sewage deposited on a beach in Cornwall was not lawful. The Council's duty, if it found there was a statutory nuisance being caused, was to serve an abatement notice.

There are two golden rules that must always be borne in mind when contemplating judicial review:

- Speed is essential – delay can be fatal to a successful challenge.

- The court will not second guess the decision-taker's decision on the merits – cases will not succeed merely because the court disagrees with the importance given to various factors in reaching a decision.

Is there a decision?

When considering whether to bring a judicial review, the first step is to identify the decision that is at the heart of the challenge. This is not always as easy as it sounds. Decisions such as the resolution to grant planning permission by a local authority planning committee, or the decision by the Environment Agency to grant an integrated pollution control authorisation, are straightforward examples. Other examples would be decisions included in letters from regulators not to disclose environmental information or not to investigate a complaint about a statutory nuisance. It can often be necessary to write a further letter to decision-takers with a clear request for them to make a decision. Their failure to do so or an unacceptable stance set out in the letter can then form the basis of the challenge.

Many of the decisions that are susceptible to challenge are made under statute. It is therefore important to be aware that a decision that purports to be made under an Act of Parliament must fall within its terms. A decision made under statutory rules or regulations must be within the scope of the Act of Parliament under which those rules or regulations were made. A decision made in line with the policy of a government department or a local authority must be both within the terms of that policy and also within the scope of the Act of Parliament and/or regulations under which they were made.

Which decisions and bodies can be judicially reviewed?

Judicial review is only available to challenge certain decisions. These decisions must be made by public bodies. In the field of the environment these bodies will include:

- government ministers;

- local authorities;

- the Environment Agency;

- regulators of private utilities such as the water regulator OFWAT, the rail regulator OFRAIL and the gas regulator OFGAS;

- privatised utilities such as water companies, British Gas, etc.;

- nationalised industries such as what remains of British Rail;

- planning inspectors and others appointed by government departments.

Organisations that are considered to be clearly 'private' will not be susceptible to judicial review. This would include private and most public companies, trade associations, clubs and individuals. Not all decisions of a 'public' body are reviewable either. The decision-making function to be challenged must be a public law one.

On what grounds can a decision be judicially reviewed?

Because judicial review has grown in a piecemeal way over the last 20 years, it is not always clear exactly what the basis is for a judicial review challenge. A judicial review can only succeed if grounds for challenging the decision are identified and subsequently proved. It is essential that a clear legal basis of challenge is determined before the action is commenced. The grounds set out below have no formal basis but are generally accepted as the categories recognised by the courts. The four main grounds are:

- illegality,
- irrationality,
- procedural irregularity,
- non-compliance with European law.

Other areas on which legal challenges have succeeded include:

- breach of legitimate expectation,
- duty to give reasons.

An essential factor to remember is that a challenge will not succeed purely because the court takes a different view on the weight it would give to relevant factors in reaching a decision. If the decision-taker has a discretion in reaching its decision, except in cases of irrationality (see below), the court will not substitute its own view.

Illegality

The requirement is that a decision-maker must understand correctly the legal framework (which may include government circulars, etc.) and give effect to it. The ground of illegality can be subdivided as follows:

1 error of law – where the decision-maker misunderstands the legal basis on which the decision is to be made;

2 *ultra vires* – or beyond the powers – where the legal framework does not give the decision-maker the power either to make the decision or to make it in the way in which it has been made;

3 fettering discretion – public bodies are entitled to develop policies about how they will make decisions, but are not entitled to use the policy in a way that means they do not consider the circumstances of each application;

4 improper delegation – if the power to make a decision is given to a public body, it must not improperly delegate the decision to someone else.

In making their decisions, bodies must take account of all relevant considerations and not take account of irrelevant ones. They must not use the powers to achieve a purpose other than the purpose for which they are given the power.

Irrationality or 'Wednesbury unreasonableness'
In general, although errors of law are grounds for judicial review, errors of fact are not. In many cases, the basic grievance of those wishing to challenge decisions made by public bodies is that they do not agree with the decision. The basic principle (set out in the 1948 case of *Associated Picture Houses Ltd* v. *Wednesbury Corporation*, and therefore also known as 'Wednesbury unreasonableness') is that, provided the decision-maker has regard to all relevant considerations, the weight or emphasis given to these matters is up to the decision-maker. Challenges based on disagreements of fact and degree, judgment and discretion are unlikely to succeed. It is only in extreme circumstances, where the court can be convinced that no reasonable authority taking account of all relevant considerations could have arrived at the decision, that the ground of irrationality can succeed. In these cases the challengers are effectively showing that, given the decision, discretion could not have been properly exercised. Successful challenge on this ground alone is *very difficult* to achieve.

Procedural irregularity

The ground of procedural irregularity deals with issues concerning how the decision is reached rather than the substance of the decision itself. In broad terms the requirement on public bodies is that in making decisions they act fairly and comply with the rules of natural justice. Procedural irregularity would include breaches of formal rules laid down by Parliament or from other relevant sources, including standards imposed by the courts themselves.

Embodied in this approach are two main areas: impartiality of decision-makers and the right to a fair hearing. The right to a fair hearing is particularly important. For example, even if there is no formal right for parties to be heard before a decision is made, if the developer is afforded an opportunity to present the case for a development it would be unfair for objectors not to be given an opportunity to present their case.

Non-compliance with European law

An important ground of challenge to an action or decision of a public body arises in circumstances where it is in breach of European Community law. In such circumstances, even Acts of Parliament can be reviewed. This ground is really just a sub-category of the *ultra vires* category of illegality referred to above.

Duty to give reasons

Increasingly it is being held by the courts that, as part of a requirement for openness, reasons should be given for decisions by public bodies. Whether such a requirement will be imposed by the courts is somewhat unpredictable. It is particularly likely to be imposed in cases where the decision is extremely important to those affected and also where it is necessary to explain what would otherwise seem to be an unexpected decision.

Legitimate expectation

In certain cases where no formal rights to be consulted exist, the conduct of the public body concerned may create a 'legitimate expectation' that the person will be heard. This approach is developing to include circumstances where a public body has given assurances that a power would be exercised in a certain way. If subsequently it does not, good reasons will have to be given for the change of approach.

Who can bring a judicial review?

There are two major considerations in determining who can seek a judicial review. These issues are known in legal jargon as *standing* and *capacity*.

Standing

A person seeking a judicial review must have a 'standing' or 'sufficient interest' in the matter to which the decision relates in order to bring an action. This issue is decided in the circumstances of each case and as a result it is difficult to be precise about who will and who will not qualify. Until recently, the courts seemed to be applying this rule quite strictly, but over the last few years a much more liberal approach has been taken, with the courts now being less likely to refuse to proceed with a judicial review on this basis.

Clearly, if the decision is precisely aimed at certain people, they will have standing to challenge the decision. In most environmental cases the link between the decision and the applicants is not so clear. This means that it is not always easy to find an appropriate person to bring the action.

In the following cases the courts accepted that the applicants had standing:

- a local recreational fisherman who challenged the grant of a waste-management licence because of concerns about potential contamination of the water in which he fished;

- users of a beach who challenged a failure by a local authority to serve an abatement notice on a water company discharging untreated waste so as to cause a nuisance on the beach;

- residents and business people in a local community who objected to the grant of planning permission for a quarry in their village;

- Greenpeace and Friends of the Earth generally challenging decisions in environmental matters.

The issue of standing frequently links up with eligibility for legal aid. Since it is impossible for legal aid to be granted other than to individuals, if a legal aid application is to be submitted it must be on behalf of a person who has a sufficient interest in the case.

Capacity
A question frequently related to standing is the matter of capacity. This concerns whether the potential applicant is a 'legal person' able to bring an action in the courts. The definition of a 'legal person' – an entity capable of enjoying and being subject to legal rights and duties – clearly includes both individuals and companies. The question is much less clear when dealing with other organisations in which people gather together, such as an informal local group. These are known under the umbrella term of 'unincorporated associations'. They can range from long-standing organisations, with a constitution but no formal legal framework or company registration, to a local informal pressure group that has been set up to deal with a single local environmental issue.

The case law is unclear about whether such unincorporated associations have the legal capacity to bring judicial reviews, but there certainly have been cases where applicants have been unable to proceed with a case because of this issue of capacity. It is therefore very risky to bring a judicial review in the name of a local group, whether a recent *ad hoc* group or one with a more formal constitution. Clearly this does not apply to groups such as Greenpeace, Friends of the Earth, or the Council for the Protection of Rural England, which all have a company framework.

This can cause difficulty. It is understandable that a local group, set up to take environmental action, which has fund-raised to run that action, would want to run it in its own name. A safe way to overcome this difficulty is to have at least two applicants. The unincorporated association can be put down as the first applicant, followed by an individual who definitely has standing as the second applicant.

Since the second applicant has capacity there is no point in the respondent authority (which is the subject of the action) challenging the capacity of the first applicant, as the case would continue in any event. The case then continues to be known by the name of the first applicant – the unincorporated association.

The need for speed

A vital issue that needs to be understood by any individual or group considering bringing a judicial review is timing. The rule is that an application for judicial review must be brought promptly and in any event within three months from the date when the grounds for the application first arose. The mistake made by many potential applicants is to consider that this means there is a period of three months within which a judicial review can be sought. This is not the case. The primary duty is that the application must be made promptly.

Many cases fail on grounds of delay. For example, in *R. v. Swale Borough Council*, ex parte *Royal Society for the Protection of Birds* (1990), the precursor case to the one cited above, the RSPB challenged the grant of planning permission for land reclamation. Although it was accepted that there had been a breach of a legitimate expectation to be consulted, it was decided that because of delay no relief would be granted.

What exactly is meant by the term 'prompt' is not easy to judge. Broadly, however, it means that, once campaigners are aware that there is a matter they may wish to challenge, they must take steps to move the issue along as fast as possible. Delay is particularly important in cases where a third party, such as the person whose licence is to be challenged, is 'prejudiced', or disadvantaged by prejudice. Take the example of a company that has been granted an integrated pollution control authorisation and subsequently spends a large sum of money implementing the authorisation. It will then be argued by the company that it has suffered substantial prejudice because of the delay in bringing the action. In such cases, it makes good sense to ensure that affected third parties are aware that a judicial review is being contemplated.

If the court accepts that there is a 'good reason' for delay, it is even possible to extend the period beyond three months. Again, since this is a matter for the court to decide, it is difficult to be certain what will or will not constitute a good reason. However, it is likely to be particularly difficult to persuade the court to agree to extend time in cases where a third party has been prejudiced by the delay.

What orders can a judge make?

It is important to be clear from the start what the effect of a successful judicial review challenge could be. It must be noted that only rarely will the court finally resolve the matter. Most frequently, a victory in court will result in the matter being sent back to the public body to determine.

There are six orders the court can make, but it must be remembered that any relief is discretionary. This means that, even if the public body is shown to have acted unlawfully, the court may not order the steps necessary to overcome the unlawful action. The principal reasons for refusing relief are concerned with the conduct of the applicant, particularly delay, and the impact on an innocent third party who has acted in reliance on the decision. This again highlights why open and speedy action by the applicant is essential.

The six orders that the court can make are: *certiorari, mandamus,* prohibition, an injunction, a declaration and damages.

- *Certiorari* has the effect of quashing the decision under challenge. No new decision is substituted but the body must then reconsider its decision in the light of clarification given by the court.

- *Mandamus* compels the public body to carry out its duty accordingly to law. This order is sought in cases where the body is failing in some way to perform a duty or exercise a discretion. Frequently applicants seek orders of *certiorari* to quash a decision and also *mandamus* to force the body to retake its decision. Courts are often reluctant to order *mandamus* to compel public bodies to carry out their functions in prescribed ways.

- Prohibition is used where the public body is anticipated to make an unlawful decision. This order prevents it from doing so.

- An injunction orders a body to do something or not to do something. Failure to comply with an injunction is a contempt of court. The injunction is much more often sought as an interim remedy to preserve the status quo pending the outcome of the case.

- A declaration involves a statement by the court of the correct legal interpretation of the matter before the court. It does not have the effect of imposing any action on the public body but clarifies the approach to be taken.

- Damages are infrequently awarded in judicial review. There is no right to damages for public law breaches. The only basis on which damages may be sought is if a tort is proved in addition to the public law matter.

PRE-REVIEW CONSIDERATIONS

As stated earlier, before a judicial review can be considered the decision under challenge must be clearly identified. Frequently, the behaviour of the public body under challenge is contained in a number of documents and exchanges of correspondence. It is necessary to look through all these documents to find exactly what decision is at the root of the problem. Often, even when points are clearly put to the decision-maker, the response is woolly and vague. It is essential to remember that, once the case comes to court, the judge will need to know exactly what is being challenged.

Public bodies can only act in ways in which they are authorised by law. In relation to the environment, this is based on legislation derived from Acts of Parliament and from European Union legislation. The details can be found either in the Act of Parliament itself – for example, the Environmental Protection Act 1990 – or in delegated legislation, such as the Waste Management Licensing Regulations 1994.

In the environment field, the body under challenge will have been set up by statute. For example, the Environment Agency was set up by the Environment Act 1995. Such bodies are only able to carry out the functions given to them under legislation. This can extend to other actions that are necessary for them to take to be able to carry out those statutory functions.

These obligations of a public body are either formulated as duties, which the public body *must* carry out, or powers, which it *may* carry out. Frequently the legislation setting out these obligations embodies constraints on the ways in which they should be exercised, and it is necessary to examine the statutory provisions to establish the exact nature of the obligation. Failure by the body to operate within this framework is unlawful and will form the basis of a judicial review challenge. EU law is frequently very important in the environmental field. All domestic legal duties and powers must be carried out in line with any relevant EU legislation.

Often before the decision under challenge can be clearly identified, further information is required. Frequently, the public body is under a duty to disclose information.

Letter before action

Once it is clear what the conduct under challenge will be, it is essential to send a letter to the public body, putting it on notice of the basis of the case being made. The letter should:

• provide the background facts;

• identify the decision under challenge;

• explain the basis on which the decision is considered to be unlawful;

- request a review of the decision (except in cases where the authority does not have the power to review the decision, such as a planning case where planning permission has already been granted);

- request reasons for the decision taken;

- stipulate a limited period for reply;

- explain that an application for leave for judicial review may be made after the specified date without further notice.

It may well be clear that the public body will not change its decision. Even in these cases the letter should be sent. It is also essential to provide a clear statement of why the decision is considered to be unlawful. This letter will form an important part of the case before the court. From it (and any response received from the respondent public body) the judge will be able to see the bare bones of the applicant's case and the response to it.

A copy of the letter should also be sent to any third party affected by the decision under challenge – for example, in the case of a judicial review of the grant of a waste-management licence, the company applying for the licence.

The approach to judicial review

It is important to understand a couple of critical differences between judicial review and many other forms of litigation. They are fundamental to the conduct of a judicial review.

Full disclosure
Many people who have experience of litigation (even just from television) have the idea that it is worth holding details of the case against the other side up their sleeve to be disclosed at a later date. This is not permissible in judicial review, and may be heavily criticised by the court, even resulting in relief being withheld in the event of success. The reason for this is that the case against the

public body will be that it failed to use the proper approach in coming to its decision. It must be borne in mind that a frequent outcome on a successful judicial review will be that the public body will be ordered to go away and reconsider its decision using the correct approach or taking into account relevant factors. If there are matters that have not been put to them where this could readily have taken place, the public body may well say that it would have acted differently if they had been brought to its attention and nothing will have been gained.

Avoid technicality
Judicial review is refreshingly free from the procedural wrangles that take up much of the court's time in other forms of litigation. In general terms, the court is concerned with the broader principles of justice in the case rather than whether or not a piece of evidence has been filed precisely at the right time. It is always wise for the applicant not to rely too heavily, therefore, on the fact that the respondent public body has done something wrong if ultimately this wrong has been corrected.

Are lawyers essential?

Because the procedure involved in judicial review is reasonably straightforward, it may be tempting to think that a case can be taken without instructing lawyers. This can be difficult. The case will be conducted in the High Court, which is very much the preserve of lawyers. Also, the legal concepts and the rules that can and can't be broken are very specialised and it is highly advisable, therefore, to seek legal advice early on.

A limited number of solicitors and barristers specialise in this kind of work. Since most lawyers have very limited experience of judicial review, it makes a great deal of sense to approach someone who deals with such work on a regular basis. Because of the short time limits involved in judicial review, approaching someone less experienced may result in poor advice and delays, which can be fatal to the success of the action. It is of the utmost importance to seek initial expert legal advice as soon as a judicial review challenge is contemplated.

How much will it cost?

The cost of bringing a case is a major factor to be considered early on by anyone contemplating a judicial review challenge. One of the first questions the lawyer should be asked is how much it will cost. It is very difficult to give an accurate response to this question. There are, however, a number of points to understand.

The loser pays the costs

The ultimate cost will be largely determined by whether the case is successful or not. Although the judge hearing the case has a wide discretion about the award of costs, the losing party will generally have to pay the other side's costs. So if the applicant wins, the costs bill of the applicant is likely to be small. If the applicant loses, it is likely that the bill will include the respondent's costs. Although others may be joined in the proceedings as second respondent, etc., it is not normal that the applicant will have to pay more than one set of costs, even if unsuccessful.

So in calculating the possible exposure to costs, the costs of both the applicant and the other side must be considered. The decision to proceed can, however, be taken in stages, and an applicant can pull out at any stage, although the respondent's costs to that point will normally have to be paid.

The applicant's costs

Calculating the applicant's costs is the easier part. It would be hoped that most solicitors who act for applicants would be willing to give an initial assessment of the merits of the proposed case free of charge, or at least for a reduced fee. At that stage, it is suggested that the solicitor be asked to give details not only of the likely success of the case but also of the likely costs of conducting the case. It is a requirement imposed on solicitors by the Law Society to provide their clients with such an estimate.

The costs charged by solicitors and barristers are largely negotiable, so it is important to clarify the charging basis from the start. Much of the cost of conducting the applicant's case is incurred in the initial preparation, establishing the grounds of challenge and collecting information to support the case. The overall costs of the case will therefore depend to a great extent on how much of this work needs to be done.

Usually a solicitor will charge the client on a quarterly basis or in advance. A barrister will be required to draft the application for leave and to appear in court on any hearings.

After the leave application, the costs will depend to a large extent on how the case proceeds and how it is conducted by the respondents. At least clients can keep a track of the costs on their own side as they grow. Decisions can also be made to limit costs – for example, by restricting the amount of preparation carried out or perhaps by using a more junior barrister for hearings.

The respondent's costs

No matter how confident the applicant is, it is essential to remember the potential liability for the respondent's costs. In the event of losing the case, it is likely that there will be liability for the other side's costs. In a recent case, however, Greenpeace was successful in resisting a costs order because the judge accepted that it was a case that raised issues of substantial public interest. Whether such an application would be successful would not be known until the end of the case. Here there is much less certainty about the size of the bill. Although the respondent would ultimately have to justify the level of costs to a judge in the event of a dispute, they will be able to claim any costs reasonably incurred in conducting the case. This means that if they employ expensive city solicitors and senior barristers and even Queen's Counsel the costs will usually be recoverable. Clearly, for bodies with large budgets, these kinds of legal costs are a routine expense. They are often aware that by running up large bills they also intimidate an applicant worried about the costs of losing.

As a rule, the work involved in preparing a case for a respondent is much more limited. They tend to have more of the information about the case to hand and also have the more limited task of responding to issues raised against them. As a rough rule of thumb it is often possible to assume that the respondent's costs will be approximately the same as the applicant's up to the time of trial. This offsets their larger hourly rate and brief fees against the fact that less work will need to have been done.

At the trial this might be different. If a QC is used the respondent's costs may well exceed those of the applicant.

How can the costs be paid?

The issue of funding cases in general is dealt with in Chapter 3. The points here apply specifically to judicial reviews.

Legal aid

Individual applicants of very limited means can seek legal aid for judicial review actions (see Chapter 3). There are a number of considerations that regularly arise with legal aid applications for judicial review. They include questions such as whether the potential costs of the action are justified by the benefit to the applicant of running the case and whether other people stand to benefit from the application. Groups cannot obtain legal aid and an assertion that the case is of general public importance is unlikely to assist.

Seeking legal aid for applicants for environmental judicial review is a very specialised field and without question a well-prepared application stands a much greater chance of success. It is therefore very important to seek early advice from experienced solicitors.

Fund-raising

Many judicial reviews are run on behalf of applicants who fund-raise among their members and wider supporters. Although this is difficult because of the potentially large amounts of costs involved, working with solicitors experienced in this field can limit the costs and the risks. Schemes do exist for the provision of low-cost and, in some cases, free services of lawyers.

How long will it take?

This is another question regularly asked by those wishing to challenge the decisions of public bodies by way of judicial review. Again, there is no straightforward answer. The courts are very busy and it can take up to 18 months for a case to come to trial.

This can be of particular concern to either or both parties, depending on whether the offending decision is implemented pending the outcome of the case. For example, where the construction of a bypass is suspended until the hearing of a judicial review application, the promoter of the road would be most unhappy. In other cases where the action complained of continues until trial, the applicant will be justifiably unhappy. The courts are renowned for their pragmatism, and if a large amount of money has been spent in implementing a decision it is much more unlikely that it will be overturned at a later stage.

The court is willing, where it is convinced of the need for speed, to 'expedite' the proceedings, and in some cases the full hearing has been held within a matter of days of lodging an application for leave.

HOW TO START A JUDICIAL REVIEW

Application for leave

Before a judicial review can be commenced, leave must be sought from the court. The justification for this procedure is to weed out cases with no prospect of success. The main document required to apply for leave for judicial review is called a Form 86A. This form must contain:

- the name, address and description of the applicant(s)

- details of the decision under challenge

- the relief sought and the grounds on which relief is sought

- the name and address of the applicant's solicitors (if any)

- the applicant's address for service

- the reasons for delay (if any) in making the application

- whether an oral or paper application is made (see page 206 below).

In addition, an affidavit must be produced, verifying the facts included in the Form 86A. Sometimes it need do no more than that. In other cases more factual detail is set out in the affidavit, giving the background to the case. The affidavit can be produced either by the applicant or his/her solicitor. Because the application for leave is made *ex parte* (this means that the judge only has the applicant's case when considering the application), it is essential that all relevant facts are disclosed – even those that undermine the applicant's case. Failure to do this risks providing grounds for the respondent to seek to have leave set aside at a later date.

A paginated and indexed bundle must be produced indicating which pages or parts of pages are essential reading for the judge. An indexed and paginated bundle of legislative material must also be produced.

Paper or oral applications

The application for leave can be dealt with in one of two ways – orally in court or where the judge considers the papers without oral submissions.

Oral applications
If an oral application is requested by the applicant, the court will fix a hearing date with the barrister's clerk, which is usually a matter of a few weeks later. At the hearing, the applicant's barrister will outline the case to a single judge at the Royal Courts of Justice in the Strand, London. Usually the judge will have read the papers beforehand, so no great detail will be required in the presentation. No witnesses are required to give evidence and it is unnecessary for the applicant to attend. If the judge decides to grant leave, other issues, such as application for interim relief and special directions such as expedition, will be dealt with.

Paper applications
In the case of a paper application, or one where an oral application is not specifically requested, the bundle of papers will be passed to

the judge for consideration. Again, this can take any time from a few weeks to several months.

If the judge considers that the respondent should be heard before a decision is made about granting leave, he/she may direct that there be an oral hearing giving notice of it to the respondent. Obtaining leave may be a big tactical advantage – the respondent may think again.

Oral versus paper applications

There are three advantages in choosing a paper application. The first is that it keeps the cost down, since it will not be necessary to pay lawyers to attend court. The second is that, if the application is refused, it can be renewed orally, giving a second bite at the cherry. However, an oral hearing will be essential if the case is urgent, since it can be arranged immediately, if necessary. Also, in complex or high-profile cases, it may be considered preferable to have an oral hearing if it is anticipated that a judge is unlikely to grant leave on the papers without hearing arguments. The third advantage is that if a paper application is made, the respondent and third parties may not be mobilised to respond to the case.

Notifying the respondent and third parties

Although in theory the application for leave is generally made *ex parte* with only the applicant attending and without notifying the respondent, in practice this is less and less common. The respondent and third parties will almost always be aware of the application from the letter before action. Increasingly frequently, a set of the papers will be sent to the respondent and any directly affected third parties when they are lodged at the Crown Office, whether a paper or an oral application is sought.

Although the traditional view is that if the respondent is represented at a leave hearing the judge may be persuaded not to grant leave, it must be remembered that the test for leave is that there is 'an arguable case'. This test has been defined by the courts in various ways, including that leave should be granted if the judge is satisfied that there is an issue, or that leave should be refused if there is no sensible prospect of success. If the respondent attends the hearing and is unable to deliver a knockout blow, leave will almost definitely be granted. If the respondent is not present, the judge is more likely to consider what possible arguments could be made on behalf of the respondent and may be less willing to grant leave. Clearly, also, if the respondent is aware that leave is being sought but does not seek to resist the application, it can be inferred that the grant of leave is not strongly resisted.

Although the requirement for leave can be seen as an additional hurdle to be overcome, it is also possible for it to be used to the applicant's advantage. This is particularly the case if the respondent has contested the application. The grant of leave can then be publicised as a victory for the applicant. Bearing in mind that it is only necessary to show that the applicant has 'an arguable case', this may be achievable in a case even where a final victory at the full hearing is unlikely. Frequently, substantial media coverage for a successful leave application can be achieved to the advantage of the applicant.

Interim relief and special directions

In order for the judicial review to be effective, it is often essential that the respondent or a third party be prevented from acting in reliance on the alleged unlawful decision until the case has been resolved. This is clearly true where the actions proposed are irreversible – for example, planning permission to develop a site of special scientific importance. It is also very important where it is intended to proceed with steps that will involve substantial costs – for example, where expensive new plant or machinery is to be installed to give effect to an integrated pollution control authorisation under Part 1 of the Environmental Protection Act 1990. Because of the pragmatism of the courts, and the discretionary nature of any remedy, it is unlikely, even if the decision is found to have been unlawful at the final hearing, that an order will be made to undo such costly steps.

In such cases it is essential to seek a court order (in the absence of the agreement of the respondent or third party not to proceed) to prevent such action pending the outcome of the case. The court has the power to order a stay or an interim injunction at any stage before the final order following the full hearing. The application is frequently most effectively combined with the application for leave.

A stay is an order that the decision that is the subject of the challenge should not progress or come into effect until the outcome of the judicial review is decided. In making such an order, a cross-undertaking in damages may be required by the judge from the applicant (see below). Where other interim relief is needed, it is possible to seek an injunction – usually to prevent the public body from acting in a certain way, but sometimes requiring positive action.

The court asks whether 'the balance of convenience' justifies making the order. It must be said, however, that the court considers such action, before the case has been fully argued, to be a draconian step not entered into lightly.

Cross-undertaking as to damages

The major difficulty for the applicant, which arises in seeking an interim injunction or stay in cases where the order would have financial consequences for the respondent or a third party, is a requirement to a give a cross-undertaking as to damages. This means that the court requires the applicant to undertake to pay the costs incurred by the respondent or third party as a result of the order, in the event that the case is not upheld at the final hearing. Since the possible costs of such an undertaking could amount to many thousands of pounds, or even millions, it is unlikely that any applicant could afford to give one.

The court has a discretion not to seek an undertaking and such a course is usually proposed on behalf of the applicant. Even if it would be impossible to give such an undertaking, it may still be appropriate for an applicant to seek a stay or injunction. However, both Greenpeace, in *R. v. Inspector of Pollution* ex parte *Greenpeace Ltd* (1994), and the RSPB, in *R. v. Secretary of State for the Environment* ex parte

RSPB (1995), failed to get a stay because they could not give a cross-undertaking. If it becomes clear that the stay or the injunction would have been made but for the failure to give the undertaking this can be made the subject of press attention.

Expedition

A possible alternative to seeking interim relief, which may be acceptable to the respondents, is to expedite the substantive hearing of the judicial review. In appropriate cases, the court has been known to order a full hearing of a judicial review within a matter of days of leave having been granted.

Discovery

If further information and/or documentation within the possession of the respondent is required by the applicant to be able to substantiate the case, it may be possible to seek an order for discovery from the court. Again, this may well be sought at a successful leave application. Unlike in other civil proceedings, an order for discovery is not common in judicial review. It will be particularly difficult to obtain such an order where evidence is sought to contradict statements made on oath by the respondent. (For more on discovery, see Chapter 12, pages 303–5.)

Renewing the application in the Court of Appeal

If leave for a judicial review is refused after an oral hearing (whether or not it was a renewed paper application) a further application for leave can be made to the Court of Appeal within seven days of the hearing unless an extension has been granted by the Court of Appeal. It is worthy of note that it may be a matter of some months before this renewed application is heard. If the Court of Appeal refuses leave there is no further appeal to the House of Lords.

WHAT HAPPENS ONCE LEAVE IS GRANTED?

Notice of Motion

Once leave has been granted, the application for judicial review must then be made. The simple standard Form 86, also known as the Notice of Motion, must be completed. Where leave has been granted on the papers, a copy of Form 86 will be provided by the Crown Office, with notification of the grant of leave. Following an oral hearing, a copy of the form will be sent shortly afterwards.

The applicant must serve the completed Form 86 on 'all persons directly affected', along with a copy of the Form 86A, affidavits in support of the leave application, and the order granting leave. Unless a specific order was made by the court, the Form 86 and supporting documents must be served and then set down in the Crown Office within 14 days of the grant of leave. Setting down involves filing two copies of Form 86 and an affidavit of service at the Crown Office.

There is no definition of who is 'directly affected', but it will always include the respondent, and in environmental cases will often include the persons whose consent or permission is challenged. It is open to anyone who considers that he/she is directly affected by the application to apply to be joined as a respondent. This is a course of action that environmental groups may consider taking where a judicial review is sought by a body challenging a decision supportive to the environment. For example, if a polluter seeks to challenge a decision by a public body that is considered to be environmentally friendly, campaigners may seek to intervene on the side of the public body.

Respondent's evidence

Once the Notice of Motion has been served on the respondent and other third parties, they normally have 56 days to file evidence in reply. If the matter is urgent then this period may have been shortened by the judge granting leave, to achieve an expedited hearing.

This time limit is frequently broken by respondents. Although an applicant may well wish to resist delays by respondents in filing evidence in reply, it is worth bearing in mind that the court will rarely take any action as long as the applicant has had a reasonable period to consider the evidence before trial. Since the evidence in reply is the only material that the respondent puts before the court, it usually contains a full justification for the actions under challenge.

What does the applicant need to do?

Having received the respondent's evidence, the applicant has a duty to reconsider the case. Frequently, at that stage, the applicant will need to consider:

- amending the Form 86A,

- seeking to settle the case with the respondent,

- filing further evidence.

Amending the Form 86A
If the respondent has adequately dealt with a part of the case set out in the Form 86A, it may be sensible to delete those aspects of the case by amending the Form 86A. This will help to avoid giving the respondent an opportunity to focus at the trial on the weaker points of the applicant's case. Equally, if the respondent's evidence discloses new grounds of challenge, it may be appropriate to seek to amend the Form 86A to include these additional points.

Settlement
Having considered the respondent's evidence, it may become clear that no case against the respondent remains. In legal aid cases, the lawyers will have no choice but to decline to proceed. In other cases, an applicant who proceeds will face the prospect of having to pay the respondent's costs.

It may be appropriate to seek to settle the case with the respondent. Frequently in these circumstances respondents will agree not to

seek their costs against the applicant. The applicant may feel understandably aggrieved that, if the matters contained in the respondent's affidavits had been disclosed earlier, proceedings would not
have been issued. Applicants may think it reasonable in these circumstances to seek to recover their costs to date from the respondents. Unfortunately, this is difficult to achieve. Unless the respondent agrees, it would be necessary for the court to hear a substantial
part of the case to clarify whether a case existed before the disclosure
of that evidence by the respondent, but the court is generally reluctant to hear a case just to determine the matter of costs.

Further evidence

The applicant frequently wishes to add further evidence to be
relied on at trial. This may be either evidence that was still in the
course of preparation when the application for leave was lodged or
matters raised in response to the respondent's evidence. Again, the
court's approach to further evidence is very flexible. In practice,
although the leave of the court is required to admit such evidence,
it will be granted, provided the other parties have had an
opportunity to consider the evidence before trial.

In fact, it has become a frequent tactic used by respondents to file a
substantial affidavit with new material in the days immediately before trial. The court is likely to take a flexible approach and will be
reluctant to exclude evidence that is of relevance to reaching the
correct result, purely because it was filed at a late date. The only
options available to the applicant in these circumstances are either to
attempt to produce a hasty reply or to seek to adjourn the trial. This
latter course of action is often very unattractive to an applicant who
experienced substantial delays in obtaining the trial date.

Preparation for trial

Once the time limits for filing evidence, and so on, have expired, the
applicant's solicitor will receive a letter from the Crown Office advising that the case has 'entered Part B of the list'. This means that the
case is considered by the court to be ready to be heard. A copy of the
letter must be sent by the applicant to all other parties. Although the

JUDICIAL REVIEW 213

tone of the letter indicates that the trial could almost take place immediately, this is rarely the case. If substantial extra time is required – for example, to file further evidence – it may be necessary to consider applying to have the case stood out of the list. Usually this is not necessary and it would be more appropriate to ask the barrister's clerk to arrange a hearing date, in liaison with the respondent's barrister's clerks, that allows enough time for additional evidence to be filed. Having a hearing date tends to help to focus the minds of the parties on completing their evidence.

Paginated bundle

It is the duty of the applicant's solicitor to prepare an indexed and paginated bundle of documents for the court. The applicant's solicitor would usually send a draft index, including the leave bundle, any orders, including the order granting leave, and any further evidence filed, to the respondents for their agreement.

The trial

The trial will usually be heard by a single High Court judge, at the Royal Courts of Justice, unless it is particularly important or complex (or a criminal matter), when it may be heard by two judges constituting a Divisional Court. The hearing is held in open court, so the press and members of the public may attend.

The applicant's barrister presents the case outlining the main legal issues and the facts underpinning the action. The judge will then be referred to important parts of the evidence in the bundle, usually following the approach set out in a skeleton argument, prepared by the barrister and lodged at court prior to the hearing date. The degree of detail considered will depend on whether or not the judge seems to have read the evidence thoroughly. The barrister will then make legal submissions. The respondent's barrister will then refer to the evidence on which the respondent's case depends, followed by submissions on the law. If there is more than one respondent represented in court, each of them will go through this process in turn. The applicant's barrister will then have an opportunity to respond to any points raised by the opposition.

As has been stated, oral evidence is very rarely given in judicial reviews. Because factual details seldom form the basis of such an action, disputes in relation to them are not usually of central importance. However, in cases where there are fundamental disagreements of fact that go to the heart of the case, an order can be (but rarely is) made for witnesses to be required to give evidence to allow cross-examination. This usual absence of oral evidence means that the case tends to move quickly and can be rather dry and difficult to follow for the lay person. At times the proceedings seem like a debate or discussion between the judge and whichever barrister is standing, and it is surprising how often the discussion focuses on points that the applicant never even considered when initially bringing the case.

Usually, the judge will at least adjourn for a short period in order to collect his/her thoughts. Judgment may be given *ex tempore* or 'off the cuff'. In more complex cases, the judgment is reserved and is handed down at a later date. In such cases a confidential copy of the draft judgment will usually be provided to the parties before it is read out in open court. This has the advantage of giving the parties an opportunity to consider which orders to ask the judge to make, including orders for costs, and to consider how to deal with the media.

Appeals

Any of the parties (including third parties served as directly affected) may wish to appeal against the judgment. It may be that the applicant wins on certain points but is dissatisfied with others and/or that one or more of the respondents might wish to appeal. Leave is required for such an appeal. The application for leave to appeal must first be made to the High Court either when judgment is given or within four weeks thereafter. If leave is refused then a further application can be made to the Court of Appeal for leave.

CONCLUSION

Judicial review is a very useful weapon in the hands of those opposing a particular development, in that it highlights the issue to the public at large and can greatly embarrass decision-makers and force them into reversing a decision. However, a decision-making body with a modicum of sense can usually avoid a decision being reviewable. Even where a court order in favour of the applicants is made, it is still open to the decision-makers to come to the same view, but this time doing it in a proper manner. A victory can therefore be Pyrrhic rather than enduring.

Whether it is worth taking this route depends very much on the circumstances of each case. In general, however, it is surprising how often decision-makers do not follow proper processes in coming to a decision, and so make them reviewable.

10 Public Protest

This chapter explores the use of the law when people become involved in public protest and looks at why the right to protest is so important. It provides practical tips on how to organise a successful protest. It looks at the way the law restricts protest, identifies the wide range of different criminal offences and explains what happens after somebody is arrested for one of these offences.

It is difficult to overstate the importance of being able to protest, since this right allows us to defend and extend all our other rights.

It is sometimes easy to forget that public protest has a long and honourable tradition in this country. For example, the right to vote and the right to join a trade union were only won after prolonged and acrimonious protests. More recently, public protest played an important part in past governments' decisions to remove Cruise and Pershing missiles from this country, to abolish the Poll Tax and to severely curtail their road-building programmes.

Internationally, public protest led to Shell's decision not to dispose of the Brent Spar oil platform into the North Sea, and to the decision by the French government to stop its nuclear weapons testing earlier than it had originally planned.

There has undoubtedly been an enormous growth in the field of environmental activism over the last four or five years. The drive for this has come less from national groups – although they have certainly been involved at times – and more from new, local grassroots groups, often springing up with loose affiliations, little organisational structure and no common objectives other than to bring demonstrators together at a particular place and at a particular time.

The almost cult status afforded to 'Swampy' and his fellow tunnellers during their exploits opposing the Newbury Bypass and the second runway at Manchester Airport is a further indication that the public has a natural empathy with this sort of demonstration and with the passive but obstructive resistance that has been shown in these protests.

Public protest covers a range of different activities, including marches, demonstrations, mass trespasses, meetings, pickets, leafleting and squats. The reasons for these protests also vary greatly, from addressing single local issues to advancing national political causes.

In many cases, protests have a number of different goals, from the very specific to the far more general. A demonstration may have the important specific goal of preventing a particular woodland from being destroyed to make way for a proposed road. It may also have the wider aim of publicising the problems with the government's national transport policy and its failure to protect the countryside.

Public protests are often an effective way of putting pressure on people, usually politicians or corporations, to make decisions that they would otherwise rather avoid. The initial reaction of such people is often to ignore or ridicule the protest. Their next reaction may then be to keep their heads down and try to ride out the storm. However, concerted protest can succeed in producing the desired changes, even if the body making the change does not have the grace to admit that it has bowed to pressure but instead gives its action a polished public relations spin.

Set out below are the primary offences that protesters can be charged with. The array is bewildering, but the reality is that protesters are rarely imprisoned, except perhaps overnight, unless there have been acts of violence committed. Demonstrators should not be put off by this list as the politics of these occasions usually mean that the police will not be very enthusiastic about locking up large numbers of people, perhaps for fear of them becoming martyrs to the environmental cause.

The possible defences have not been set out, largely because there are so many that it would be impossible to cover them without the book being swamped by this chapter. Further information on criminal prosecutions is given in Chapter 16.

GENERAL PRINCIPLES OF PUBLIC PROTEST LAW

International law, most notably the European Convention on Human Rights, recognises the importance of the right to peaceful protest and that this right must be protected wherever possible (see Chapter 6 for further information on European law). Similarly, many countries have positive statements guaranteeing the right to peaceful protest.

There is currently no similar positive statement of this right in UK law, although the new Labour government has promised to incorporate the European Convention on Human Rights into domestic law. Quite how this will be done and what its effects will be are not yet clear. At the moment, the UK's approach is essentially negative, with a person being free to do anything that does not break the law. While this may sound reassuring, the freedom is largely illusory, since there are a large number of offences directed at preventing public inconvenience and disruption to the life of the community and to preserving order. Since inconvenience and disruption is often either the express purpose or an unavoidable consequence of protest, this means that many protests come into conflict with the law.

Because of the absence of provisions safeguarding the right to protest, the UK's approach is dangerously one-sided. This unbalanced approach has, in recent years, tilted even more towards the maintenance of order with the introduction of a range of new provisions that further limit public protest.

In essence, there are two sources of national law in the UK. First, there is law made in Parliament by the passing of statutes, referred to as Acts. The main Acts covering public protest are the Public Order Act 1986 and the Criminal Justice and Public Order Act 1994. Second, there is the 'common law', which is derived from the decisions of judges made over the years in many, many cases. An example of this judge-made law is the power of the police to deal with breaches of the peace.

Often the two types of law go hand in hand, with particular words or phrases in Acts being interpreted and defined by judges in particular cases. For example, the words 'threatening', 'abusive' and 'insulting', which are all contained in the Public Order Act 1986, have been interpreted by judges in particular cases.

There is a bewildering array of statute law covering public protest. Many of the provisions were passed with a particular perceived problem in mind and with little regard to the wider consequences. This piecemeal approach has led to considerable inconsistency and overlap between different provisions. It is quite possible for a protester, at any one time, to be committing a number of different offences that are very similar in nature.

In addition to the range of criminal offences defined in national law, there can also be important restrictions placed on protesters by bylaws imposed by particular local authorities.

The picture is further complicated by the extremely variable way in which these provisions are enforced. There are differences in enforcement by police forces across the country, by the same police force in relation to different types of protests and even by the same police force at different times when responding to the same type of protest.

An underlying difficulty with this whole area of the law is that it is often a matter of opinion whether an offence has been committed or not. If a person throws a brick through a window then this is likely to be an offence. However, the decision whether a protester is obstructing the highway by standing on the pavement handing out leaflets is far more open to interpretation, initially by the police and subsequently by the courts. A protester may, therefore, be arrested for committing an offence that he/she does not believe has been committed.

Another major problem is that many of the provisions are preventative, in that they aim to stop outbreaks of public disorder from occurring in the first place. Again, there can be differing views about whether a particular activity is likely to cause such an outbreak.

Given that it is the police who decide whether a public disorder has occurred or is likely to occur and given the variability in the interpretation, it is not surprising that protesters can lose confidence in the impartiality of the police and consider their decisions to be motivated by baser, more political considerations.

Practical tips for organising an effective protest

Form a committee

It is often useful for protest organisers to form a committee and designate specific roles to specific people. The exact roles will obviously depend on the type of protest and the jobs to be done. It is likely, however, that most committees will require a chair, a secretary, someone to control the financing of the protest, including fund-raising, a publicity officer and a press officer. Some of these roles may be rolled into one, so that often the main two or three driving forces of the protest will take on these separate tasks, without necessarily formalising them in this way. It is not the name or position that is important but more that these tasks are thought through and divided up so there is neither duplication of work nor significant gaps. (See Chapter 2, pages 25–31, for more on forming a group.)

Decide on the type of protest

It is important to consider the aims of the protest, ranging from the very specific to the more general. Only then can the most effective form of protest be considered. The organisers should then agree a date, time, venue and location, estimate probable attendance and consider the likely disruption to the local community.

Notify the police and other bodies

Even if it is not a legal requirement, it is often useful to liaise with the police and, in particular, to discuss the details of the proposed protest. In addition, the permission of others, such as the landlord or local authority, may be required. On some occasions, such as organising a mass trespass, it would clearly not be appropriate to give advanced notification.

Maximise publicity

The organisers should consider ways to publicise the protest and should draw up a media list identifying journalists, their telephone and fax numbers and their deadlines. The organisers may wish to issue a press release in advance giving details of the protest. More recently, the Internet has become increasingly important as a way of publicising a protest and contacting other protesters.

Create an organisers' office
The organisers should try to arrange to have an office as close to the demonstration on the day of the protest itself, which should be staffed throughout this period. Ideally there should be two telephone lines to the office, one to speak with the police, stewards and observers and the other to speak with protesters and take press enquiries. In these days of mobile phones a direct connection between the office and those heading the protest is a distinct advantage.

Appointing stewards
If the protest involves a large number of people, stewards should be appointed to help ensure that the event runs smoothly. These stewards should wear distinctive clothing (such as bibs, arm-bands or badges) so that they can be clearly recognised by protesters and the police. Ideally, the stewards should again have access to mobile phones to ensure that they are in contact with the organisers' office and others on the protest.

Recruit independent observers
It may be useful for independent observers to be present at the protest. Their presence often tends to calm situations and inhibit excessive force by the police. These must be independent of the protest and suitably marked. If possible, observers with legal training should be used.

Ideally, observers should have camcorders or cameras in order to record events. Video evidence is becoming an increasingly important part of many court cases. Failing this, dictation machines, tape-recorders, or even a pen and paper should be used. These observers should also have mobile phones to ensure that they are in contact with the organisers' office.

Observers play a vital role in recording events. If there are any incidents, especially if protesters are arrested, a detailed note should be taken of the incident, including the numbers of the police officers involved and the names and addresses of any witnesses.

Distribute solicitors' names

It may be useful to provide protesters with leaflets identifying solicitors who are prepared to act for them if they are arrested. The organisers should have contacted solicitors, who are willing to act, in advance.

Meeting after the protest

It is important for the organisers to arrange a debriefing meeting to discuss how the protest went (and particularly any problems encountered) with stewards and observers. It may also be useful to arrange a meeting in the days following the protest for all those arrested and charged at the protest, to make sure they have all found solicitors and that all information from the observers and other protesters is passed on.

POLICE AND COURT PROCEDURE

As has already been said, there are a large number of public order offences. Details of the most important of these offences are given in the following sections of this chapter and have been divided into the more general offences, which apply to almost all forms of protest, and more specific offences, which relate to particular types of protest.

The police are specifically given the power to arrest protesters for most public order offences. However, even where they do not have specific powers, they may be able to use their general arrest powers. These general powers can be used if the person refuses to give a correct name and address or when it is necessary to prevent a breach of the peace, to stop the person injuring (or being injured by) someone else, or to stop the person damaging property or obstructing the highway.

Where a protester is arrested, that individual must be told at the time of the arrest, or as soon after the arrest as possible, the reason for the arrest. The protester should then be taken to a police station as soon as possible for the charge to be formally presented. However, if police resources are at full stretch at the protest, this may not happen until some hours afterwards.

On arrival at the police station, the person should be informed of his/her right to legal advice. The person is entitled to advice from a specifically chosen solicitor or from a so-called 'duty solicitor'. The duty solicitor scheme ensures that a solicitor is available on a 24-hour basis to attend police stations to advise suspects who are being detained (see Chapter 3, page 34). This initial advice is free of charge, irrespective of means. In addition, depending on the means of the person and the nature of the offence, a person may be eligible for criminal legal aid (see Chapter 3, page 38). The solicitor will discuss these issues during the first meeting with the client.

Generally, the police can detain a person for up to 24 hours following arrest. During this time they will interview the person detained and decide whether they have enough evidence for a charge to be laid.

If a person requests legal advice, the interview with the police should normally wait until a solicitor has arrived and the arrested individual has been given the opportunity to speak privately with the solicitor.

Changes brought in by the Criminal Justice and Public Order Act 1994 now allow the court to draw adverse inferences from a failure to answer police questions during interview or a failure to give evidence at trial. This is a worrying development, although those who support the change make the point that an adverse inference is by no means automatic and there may well be very good reasons why a defence to the charges is only given after the initial interviews. However, it certainly does mean that a potential defendant will have to think twice before refusing to respond to questions by the police. A solicitor will discuss this with the client prior to any interview by the police.

If a person is not charged, release must take place after a maximum of 24 hours' detention, although the police require authorisation to keep a person detained without charge for even this long. If a person is charged with an offence, the person may be released on police bail, which obliges the person to attend a magistrates' court on a fixed date to answer the charge. If police bail is not given, the person is detained until the next available sitting of the magistrates' court.

Magistrates, also known as justices of the peace (JPs), will usually sit in twos or threes, unless they are legally qualified. Where there is more than one magistrate on the bench, that will almost certainly mean that they are part-time volunteers, who will be assisted by a legally qualified court clerk.

For less serious offences, where the alleged offender has not been arrested, details of the offence are presented before the magistrates, who issue a summons, which is subsequently posted to the person. The summons will require the recipient to appear at the court at a specified future time.

Criminal offences are divided into summary offences (the least serious), offences tried only on indictment (the most serious) and those triable either way (middling seriousness). Most public order offences are either summary offences or offences triable either way.

Summary offences are tried in a magistrates' court by the magistrates. Offences triable on indictment are tried in the Crown Court, where the case is heard by a judge and jury. Offences triable either way can be tried in either the magistrates' court or the Crown Court, with the venue decided at a preliminary hearing before the magistrates, who come to a view as to whether the offence is sufficiently serious that it is more appropriately tried at the Crown Court. If the magistrates are willing for the offence to be tried at the magistrates' court, the defendant is given the option of venue.

As a general rule, the major advantage of the magistrates' court is that the process is cheaper and speedier and the sentences are lighter. The major advantage of the Crown Court is that the likelihood of conviction is less (since guilt is determined by a jury rather than 'battle-hardened' magistrates) and the venue provides a better forum for challenging legal issues such as the admissibility of evidence.

Whatever the final venue, nearly all cases begin in the magistrates' court. At the initial hearing, the case will usually be adjourned – that is, the primary decision-making process regarding whether the charges have been proved is put off to another day – to allow the prosecution or defence more time to prepare their case. If a case is adjourned, the court will consider whether the person should be remanded in custody or on bail. For people charged with most public order offences, bail is likely to be granted. However, the courts may impose conditions when granting bail that may include prohibiting the person from returning to the protest. Questions of legal aid may also be dealt with at the preliminary hearing.

Before a case is transferred to the Crown Court, the case is subject to committal proceedings in the magistrates' court. At these proceedings, the prosecution must show that the defendant has at least a case to answer. Often these proceedings are a mere formality and take only a few minutes. However, on other occasions, much of the evidence is heard and the proceedings can take much longer.

GENERAL PUBLIC ORDER OFFENCES

There are many criminal offences that, although not specifically related to public order, apply as much to protesters as to everybody else in society. Examples of these offences include causing criminal damage (for example, by writing graffiti), assaulting a person (for example, by kicking or punching the person) or assaulting a person occasioning actual bodily harm (for example, by bruising the person).

In addition, there are a number of offences that relate to the breakdown of public order. These general public order offences are outlined below.

Breach of the peace

The police have wide powers to stop people from causing, or being likely to cause, a breach of the peace.

What actually constitutes a breach of the peace is not always clear, although a disruption involving violence or the threat of violence seems to be needed. This suggests that mere noise, disturbance or even abuse is not enough. However, in these circumstances, the police may be able to argue that this noise, disturbance or abuse indicated to them that violence or the threat of violence (and so a breach of the peace) was likely.

If a breach of the peace has occurred or is feared, the police have a wide discretion in the action they can take. For example, they can require protesters to leave a protest, stop protesters from arriving or detain them during the protest. They can also require that the protest be conducted in a certain way by, for example, requiring protesters to remove provocative emblems.

During the miners' strike of 1984, police turned back miners at road blocks many miles from where they intended to picket. In court, the police successfully argued that they had reason to believe that the striking miners were on their way to a mass picket and that there would be a breach of the peace should they continue. Since the miners were still a considerable time and distance away from their intended picket, this seems to have extended police powers to take action even where the breach of the peace is not imminent.

Although a breach of the peace is not a criminal offence, the police have the power to arrest a protester for a breach of the peace or where they believe that the person is about to commit a breach the peace. The person may then be taken to a magistrates' court to be bound over.

Bind-over order

A person can be bound over in a magistrates' court if the police can show that a person might breach the peace in the future. Being bound over involves the person agreeing to keep the peace and/or be of good order. The person must consent to being bound over since the penalty cannot be imposed by the court without the person's consent. However, if the person refuses, a sentence of imprisonment for up to six months can be made.

In 1961 Bertrand Russell and other colleagues in the Campaign for Nuclear Disarmament (CND) were sent to prison for refusing to be bound over. More recently, protesters at the M11 Link Road were arrested for a breach of the peace and subsequently imprisoned for seven days after refusing to be bound over.

The bind-over order will be for a specified period, generally no more than a year or two, and may also specify a certain sum of money (for example £500) that is to be forfeited by that person, or a guarantor, if the person breaks the order. The court cannot impose any other conditions on the person in the bind-over order. While a person can be imprisoned for refusing to be bound over, a person cannot be imprisoned for breaking the order. The person or guarantor merely forfeits any money specified in the order.

Bind-over orders are often used against protesters who have been arrested for breaches of the peace. However, they can also be used against protesters who are in court for other reasons. For example, they have been used against hunt saboteurs appearing in court as prosecution witnesses against huntsmen who had assaulted them!

In a similar way to a breach of the peace not being a criminal offence, being bound over is not a conviction but merely a formal promise of future conduct. Unfortunately, once an order has been granted, it severely restricts a protester's ability to continue to protest. Magistrates like to use this provision since they consider that it enables them to defuse situations by giving a person a warning about their future conduct without conferring the stigma of a criminal conviction on the person.

The police also like the provision because relatively little evidence is needed to obtain a bind-over order compared to the amount of work they have to put in to gain a conviction for other public order offences. Because of this, it is often the case that while the police may have arrested somebody for a public order offence, they offer to drop the charge in exchange for the person agreeing to be bound over.

Many civil liberties campaigners are concerned that this power can lead to people being imprisoned even though they have committed no offence. In the last few years, the Law Commission added their voice to the debate, describing the power as unconstitutional and a breach of the principles of natural justice. However, no change seems imminent.

Obstruction of the police

It is an offence under section 51 of the Police Act 1964 to assault or obstruct a police officer in the execution of his/her duty.

Almost anything that makes it more difficult for the police officer to do his/her job amounts to obstruction. Disregarding an order given by a police officer on the grounds of public safety to move clear of an area or to follow a particular route could constitute obstruction. Similarly, asking police officers why they are arresting a fellow protester could be construed as interfering with the arrest and also amount to obstruction. Nonetheless, it is important to remember that there is no positive obligation to help police with their enquiries and any failure to do so cannot constitute obstruction.

This offence is widely used by the police at public protests, often in conjunction with breach of the peace. In one famous incident, a spokesperson for the National Council for Civil Liberties was addressing a group of unemployed people outside a Jobcentre. The police feared a breach of the peace (as there had been restlessness a year earlier when the same speaker made an address) and ordered her to get down off her soapbox. She refused and was arrested for obstructing the police.

While there is no specific power for the police to arrest someone for obstructing a police officer, they may be able to use their general powers of arrest if, for example, they consider that a breach of the peace is imminent or if the protester refuses to give his/her name.

The offence is tried at the magistrates' court. On conviction, the maximum penalty for an assault under these circumstances is six months' imprisonment or a fine of up to £5,000, or both. The penalty simply for obstruction is rather less than this.

Obstruction of the highway

While a person has the right to use the highway for passage along it and for purposes reasonably associated with this, it is an offence under both common law and section 137 of the Highways Act 1980 to obstruct the highway.

The highway includes not only public roads and pavements but also footpaths and bridleways, although it does not apply to private roads or other roads where the public has only limited access. The obstruction does not have to be complete and an obstruction can be caused even if there is enough room for others to pass unaffected.

In order to commit the offence, the obstruction must be 'without lawful authority or excuse'. These are ill-defined terms that allow the court to consider the exact circumstances of the obstruction and decide whether it was reasonable.

For example, juggling with firesticks and stopping vehicles at a picket line have both been considered unreasonable and thereby causing unlawful obstruction. During the News International printworkers' dispute at Wapping during the mid-1980s, the courts also found that mass picketing was an unreasonable use of the highway. On the other hand, the courts have recently ruled that assembly on the highway to protest against animal exports was a proper use. Similarly, convictions of animal rights protesters who had been handing out leaflets in a shopping centre were quashed on the basis that the police had not shown either that there had been an obstruction or that there was no lawful excuse for it.

This offence is often used to regulate the conduct of a protest – for example, by stopping people handing out leaflets, having a 'sit-down' protest or picketing on the pavement. Often an arrest is made on flimsy grounds, the person or people are removed from the scene and the charges are subsequently dropped (especially if the protester agrees to be bound over). While there is no specific power of arrest for committing this offence, the police are likely be able to use their general arrest powers, since one of the circumstances in which they can use these general powers is where it is necessary to prevent an unlawful obstruction of the highway.

The offence is tried at the magistrates' court. On conviction, the maximum penalty is a fine of £1,000.

Riot

When twelve people together use or threaten violence for common purpose which would make a reasonable person fear for his/her safety, those using the unlawful violence are, under section 1 of the Public Order Act 1986, guilty of riot.

Riot is therefore subject to a two-part test. First, twelve or more people need to use or threaten violence for a common purpose. Then, those using violence (but not those only threatening the violence) will be guilty of riot. The reasonable person fearing for his or her safety can be entirely hypothetical. The police do not need to show that such a person was present. The fear experienced by those actually present is not relevant unless it proves evidence of what the reasonable person would have felt.

What constitutes violence is defined widely to include violent conduct towards people and property. The mere presence of a large number of people chanting verbal threats does not constitute violence (although it is obviously a threat).

A person can be arrested for riot. Prosecutions for riot are made by, or with the consent of, the Director of Public Prosecutions. The offence is tried in the Crown Court. Upon conviction, the maximum penalty is ten years' imprisonment or an unlimited fine or both. Historically, there have been very few successful prosecutions for riot as it has been difficult to show that the participants had the necessary common purpose needed to obtain a conviction. While over 600 miners were charged with riot during the miners' dispute in 1984, the vast majority were either acquitted or had the charges dropped.

Violent disorder

When three or more persons together use or threaten violence, which would make a reasonable person fear for his/her safety, each of those using or threatening violence is guilty, under section 2 of the Public Order Act 1986, of violent disorder. In order for the three people to be considered to be present together, they should at least know that the others are there, although they need not be acting in concert.

The offence is similar to riot in needing a number of people to use or threaten violence and is also similar in the need for the hypothetical reasonable person to be put in fear. However, the offence is different in that only three people are needed, they need not have a common purpose and those threatening violence as well as those using violence are guilty. These differences mean that, following an outbreak of public disorder by groups of people, they are far more likely to be charged with violent disorder than riot.

A person can be arrested for violent disorder. The offence can be tried either in the magistrates' court or the Crown Court. On conviction in the magistrates' court, the maximum penalty is six months' imprisonment or a fine or both. On conviction in the Crown Court, the maximum penalty is five years' imprisonment or an unlimited fine or both.

Affray

Where a single person uses or threatens violence towards another that would make a reasonable person fear for his/her safety, this is an offence under section 3 of the Public Order Act 1986. This offence uses similar concepts to those discussed above in relation to riot and violent disorder except that the disorder is caused by an individual rather than a group. However, an important difference is that the violence or threat of violence must be directed against another person (although the test is still whether a hypothetical reasonable person would be put in fear, not the specific person against whom the violence or threats were directed). Another important difference is that what constitutes a threat of violence is more limited. It cannot be made by words alone.

Police may arrest someone they reasonably suspect to be committing the offence of affray. However, the police do not have the specific power under these provisions to arrest a person if the offence has already been committed, although they are able to use their general arrest powers if they consider it necessary to prevent further violence or damage.

The offence can be tried in either the magistrates' court or the Crown Court. On conviction in the magistrates' court, the maximum penalty is six months' imprisonment or a fine of up to £5,000 or both. On conviction in the Crown Court, the maximum penalty is three years' imprisonment or an unlimited fine or both.

Threatening behaviour

It is an offence, under section 4 of the Public Order Act 1986, for a single person to use threatening, abusive or insulting words or behaviour, or to distribute or display any writing, sign or other visible representation that is threatening, abusive or insulting towards another person, which the person intends to or is likely to cause another person to believe that immediate violence will be used, or to provoke such violence.

This provision has been widely used at demonstrations and pickets as well as to deal with football hooligans and even streakers.

What precisely constitutes 'threatening, abusive or insulting' behaviour is not very clear, although it seems that it is something more than merely being annoying. For example, when an anti-apartheid campaigner disrupted a tennis match at Wimbledon featuring a South African player, this caused considerable annoyance to the crowd, but the behaviour was not held to be insulting, let alone threatening or abusive.

The phrase 'writing, sign or visible representation' is far-reaching and covers flags, banners, badges, emblems, armbands, and T-shirts. For example, a placard of a foetus in a pool of blood being waved by an anti-abortionist at a picket of an abortion clinic was held to be abusive or insulting.

Like the offence of affray, the conduct has to be directed at a specific person or group of people (so it could not be unlawful conduct if no one was around). As with the offences of riot, violent disorder and affray, the actual result of the conduct is not important. However, unlike these offences, it is not the likely effect on the hypothetical reasonable person that is relevant but rather the intended or likely effect on the actual person targeted. Protesters must take their audience as they find them. For this reason, a fascist was found guilty for saying that Hitler was right at a rally in Trafalgar Square. His conduct was found to be insulting to Jewish people present in the audience and to be likely to provoke them into violence.

By the same argument, police officers, through the nature of their job and training, would be expected to be able to withstand a higher threshold of offensive behaviour (compared with more timid members of society) before being able to say they believed that violence was to be inflicted on them. Similarly, the standard abuse that is hurled at those who pass through a picket line would be unlikely to allow those crossing the picket line to argue that they believed that they were about to be attacked.

The fear must be of immediate violence being inflicted. An attempt was made to bring a case against the publishers of Salman Rushdie's *Satanic Verses* for distributing abusive and insulting material on the basis that it was likely to provoke unlawful violence. This failed because any violence provoked would not be immediate enough to be covered by the section.

The police can arrest someone they reasonably suspect to be committing an offence under section 4. The offence is tried in the magistrates' court. On conviction, the maximum penalty is imprisonment for up to six months or a fine of up to £5,000 or both.

Offensive or disorderly conduct

It is an offence, under section 5 of the Public Order Act 1986, if a person uses threatening, abusive or insulting words or behaves in a disorderly way or displays any writing, sign or any other visible representation within the sight or hearing of a person likely to be caused harassment, alarm or distress. The offence is widely used against protesters and is something of a catch-all provision designed to deal with lesser activities likely to cause public disorder.

There are similarities with the offence of threatening behaviour in that the behaviour must be aimed at an audience (it needs to be within the sight or earshot of a person likely to be affected, although it does not need to be specifically directed at that particular person) and the composition of the audience is important in determining whether anybody is likely to be harassed, alarmed or distressed.

There are also important differences, however, with the offence of threatening behaviour. Most importantly, a lesser degree of 'offensive' behaviour is required, since disorderly behaviour can constitute an offence. This term has been defined widely so that, for example, standing in front of a theodolite to prevent a land engineer from taking measurements was found to be disorderly behaviour which caused harassment, alarm or distress to the land engineer.

In addition, the likely effect needed on the audience is much less, as it is only necessary to show that the likely effect was to harass, alarm or distress the audience and not to cause them to fear that violence would be used or provoked. Harassment, alarm or distress do mean more than mere annoyance, disturbance or aggravation. Nonetheless, since the aim of many demonstrations is to cause a certain amount of distress, the inclusion of this ill-defined word is a cause for concern.

The police have the power to arrest any person who persists in the offence after being warned to stop. The arrest must be made by the same officer who gave the warning. The offence is tried in the magistrates' court and carries a maximum penalty of £1,000, but not imprisonment.

Civil liberties campaigners are concerned about the catch-all nature of these provisions, since terms such as 'disorderly behaviour' and 'distress' are capable of very wide interpretation. There is also concern that this can be a 'victimless crime'. The conduct may not lead to anyone being harassed, alarmed or distressed. For example, members of OutRage, the gay campaigning group, were arrested while demonstrating with banners outside Islamic meetings, even though nobody actually complained.

In recent years these provisions have often been used to deal with offences against the police themselves, with recent Home Office research suggesting that these accounted for a quarter of all cases brought under section 5. Since protests may often be rowdy and decisions taken by the police can be seen as less than even-handed, section 5 is being used as a way of enforcing respect for the police.

Intentional disorderly or offensive conduct

A new offence of causing intentional harassment, alarm or distress
was introduced by section 154 of the Criminal Justice and Public
Order Act 1994 (and inserted into the Public Order Act 1986 as
section 4A). It differs from section 5 of the Public Order Act 1986
in that the offence is only committed if the person *intends* to cause
harassment, alarm or distress, while section 5 merely requires the
offending conduct to be *likely* to cause this effect. The offence was
originally introduced to increase the penalties available for
offensive conduct with a racial motive, although there is no
reference to this in the provisions. In fact, it is now often used as
an alternative to prosecutions under section 5 since, on conviction, the maximum penalties are greater.

The police may arrest someone that they reasonably believe to
be committing this offence and, unlike section 5, no warning is
necessary. The offence is tried in a magistrates' court and the
maximum penalty is up to six months' imprisonment or a fine of
£5,000 or both.

CONTROLS ON PARTICULAR TYPES OF PROTEST

Marches

There are important restrictions on marches contained in sections
11–13 of the Public Order Act 1986. The police can arrest either
the organiser or a protester for offences under these sections and
these offences are tried in a magistrates' court. On conviction the
maximum penalty is three months' imprisonment and a fine of
either £2,500 or £1,000 depending on the offence.

Types of marches covered
These provisions use the term 'public procession' and this is
widely defined. There is no minimum number of people needed to
make up a procession, although there would obviously need to be
more than a single protester. For the procession to come within

the definition of the word 'public', it must take place in a public place. This includes all places to which the public have access, even if only upon payment of a fee, rather than just processions on the public highway. In addition, while the protesters will usually be on foot, the provisions also cover processions by car or cycle or other means of transport.

As has been explained above, the public has the right to use the public highway for movement along it. Therefore, as long as a protester is exercising his/her right to move along the highway, the march is lawful. However, if, for example, a protester deliberately stops for a lengthy period at a particular point, a prosecution could be mounted for obstruction of the highway.

Advance notice
The organisers of certain types of processions are required, under section 11 of the Public Order Act 1986, to give advance notice of the march to the police. All processions intended to support or oppose the views or actions of a person or body, to publicise a cause or campaign, or to mark or commemorate an event are covered by this requirement. In contrast, funeral processions and processions held by custom (for example on Remembrance Day or the Easter Parade) are excluded from these requirements. It seems likely, however, that any protest march will fall into at least one of the categories requiring notice.

The notice must be given in writing and delivered by hand or recorded delivery to any police station in the police force area where the procession is intended to start. It must give details of the date, start time and route of the proposed march as well as the name and address of at least one of the organisers.

There is no precise definition of who actually can be classed as an organiser of a march. However, it is likely that anyone involved in the preparation of the procession (rather than merely being a steward on the day) would be considered to be an organiser.

The police usually provide the organisers with a standard form to complete, which asks for far more information than is specified in the legal provisions.

The notice must normally be delivered at least six clear days in advance of the march. However it is accepted that some marches will be spontaneous or need to be organised at very short notice to have any effect and in these cases notice must be given as soon as practicable. It is an offence for an organiser to fail to meet the requirement to give notice or for the march not to follow the route given (unless the organiser can show that the failure was out of his/her control or with the agreement of the police).

Since most processions were notified in the past to the police anyway, many civil liberties campaigners believe that the new provisions were introduced in order to discourage marches, by making the organisers personally liable should the march not proceed as planned.

Conditions on a march

Section 12 of the Public Order Act enables the police to impose conditions on a march where they consider there is a risk of serious public disorder, serious damage to property or serious disruption to the life of the community, or that the purpose of organisers is to intimidate others. These poorly defined terms allow the police to impose conditions on many marches.

Almost by its very nature, a march is likely to cause some disruption to the community, although whether this constitutes a serious disruption is more questionable. Similarly, the protesters on marches are often trying to embarrass or annoy others not on the march. This in itself will not amount to intimidation. For example, it was found that a demonstrator shouting 'Apartheid murderers' at a visitor to a reception at South Africa House was merely trying to discomfit rather than intimidate the visitor and was, therefore, not committing an offence.

Conditions can be imposed before a march, in writing, by the chief police officer of the relevant area. They can also be given, either verbally or in writing, during a march by the most senior police officer present. These conditions must be in response to the anticipated problems and so could relate to issues such as the route, time or duration of the procession, the carrying or use of flags, banners, emblems, loudspeakers or vehicles, or to the stewarding arrangements and provision of medical treatment.

It is an offence for the organisers to fail to comply with these conditions, although it is a defence for them to show that the failure arose from circumstances beyond their control (for example, where a protester waves a flag despite the organiser having made it clear that flags were not to be taken or used on the procession). It is also an offence for other protesters knowingly to fail to comply with a condition unless the person was unaware of the condition or the failure arose from circumstances beyond the person's control.

Banning the march
Section 13 allows the police to ban a march completely if they consider that imposing conditions would not be sufficient to prevent the risk of serious public disorder. They are able to ban a class of marches (for example, all political marches) but more usually apply a blanket ban on all marches. The ban can cover all or part of a district and can last for up to three months, although most bans are for less than 30 days.

Unfortunately, these bans, even when they only apply to a class of marches, can result in other completely unobjectionable marches having to be cancelled. For example, a Friends of the Earth rally against nuclear energy had to be cancelled after a blanket ban was imposed in anticipation of a National Front demonstration.

Outside London, the chief constable for the area needs to apply to the local authority for a banning order, the Home Secretary's approval is also needed. In London, the Police Commissioner applies directly to the Home Secretary for permission, although this may well be an area that will be taken over by the proposed mayor of London. The police must have reasonable grounds for banning a march or imposing conditions. If not, it is possible to challenge, by judicial review, any decision as being unreasonable in the courts, but, in practice, courts have shown themselves to be very reluctant to overturn decisions of the police on these matters (see Chapter 9 for more on this process).

Demonstrations

The most important provisions controlling demonstrations are contained in sections 14–16 of the Public Order Act 1986. The police can arrest organisers and protesters for offences under these sections. The offences are tried in a magistrates' court. On conviction, the maximum penalty is three months' imprisonment or a fine of £2,500 or £1,000 (depending on the offence) or both.

An offence under these sections is considered more serious than obstructing the highway, where the penalty on conviction for the protester is a fine but with no possibility of imprisonment. It is therefore a common police tactic to ask protesters to move on and then arrest them under these provisions for failing to comply with the request.

Types of demonstrations

These provisions refer to 'public assemblies', which covers gatherings of 20 or more people in a public place that is at least partially open to the air. This definition, therefore, includes not only demonstrations but also large pickets, rallies, vigils and even some queues.

The requirement for the public assembly to be at least partially open to the air covers gatherings on the road or pavement, in a park or at the entrance to a particular factory or shop. However, it does not include meetings in a building, even if the doors and windows are open.

Conditions on the demonstration

Under section 14, the police can place conditions on demonstrations either before or during the demonstration. These provisions closely mirror the police powers to place conditions on marches and can be imposed on organisers of the demonstration or on the demonstrators themselves.

Conditions can be imposed by the police if they reasonably believe that the assembly may result in serious public disorder, serious damage to property or serious disruption to the life of the community, or that the purpose of the organisers is to intimidate others.

These conditions should relate to issues such as the place, duration and number of people attending the assembly. Important considerations may be the effect of increased volume of traffic, adequacy of arrangements for off-street parking, adequate sanitary arrangements, adequacy of lighting and the avoidance of overcrowding.

This is an important power since it allows the police to severely limit the effectiveness of a demonstration by, for example, moving it to a much less sensitive location or limiting the number of protesters to 20.

Overall, however, these provisions are relatively rarely used except against pickets at industrial actions. In these cases, the police will usually cite intimidation as the reason for imposing conditions on a mass picket. However, during the News International dispute at Wapping mentioned above (page 229), it was held that mere abuse, swearing and shouting did not in itself amount to intimidation.

The provisions relating to demonstrations differ in a number of respects from those relating to marches. Most importantly, there is no requirement for the organiser of a demonstration to give advance notice to the police and the police have no power to ban a demonstration. However, it is important to stress that the police can use other powers, such as those relating to breach of the peace, to ban or break up demonstrations.

While the permission of the police is not required to hold a demonstration, the permission of other people may be necessary. Obviously, if the demonstration is to be held on private land it requires the permission of the landowner or occupier (although protesters may choose not to ask for permission and deliberately trespass on the land). The permission of the local authority is likely to be necessary for a demonstration in a park or square. In addition, any large gathering may have to be licensed by the local authority in relation to public health and environmental health matters.

Meetings

Organising a public meeting

Any meeting open to the public, even if it is held on private property or where payment is required for entry, is a public meeting.

Anybody wishing to hire a private venue (for example, a room in a pub) for the meeting, must comply with the terms and conditions in the hire agreement (for example, fire and safety).

However, the law does not give a person a right to be able to hire a particular venue. Therefore, if a local authority or private landlord refuses to take a booking from a particular group there is little that can be done.

Stewards can be used at meetings to assist entrance and exit, maintain order and eject hecklers.

Protesting at a public meeting
Any attempt by protesters to disrupt a public meeting is an offence under the Public Meeting Act 1908. This offence is tried in a magistrates' court, and on conviction the maximum penalty is six months' imprisonment and a fine of £5,000. While the police do not have a specific power to arrest somebody for an offence under this provision, they may be able to use their general powers of arrest in order, for example, to prevent a breach of the peace.

Private meetings
Any meeting at which the public are not free to attend is a private meeting, even if it is held in a public building (such as the town hall). These meetings are governed by the rules of the particular organisation holding the meeting, as well as any conditions relating to the hire of the premises themselves.

The police have no right to enter a private meeting unless they fear that a breach of the peace is imminent or they are present in order to prevent a crime.

Pickets

There are no specific provisions dealing with picketing in situations unrelated to an industrial dispute, but other provisions mentioned in this chapter (such as breach of the peace, obstruction of the highway and, if there are more than 20 people present, controls over public assemblies) are used in relation to picketing.

One important recent development is a government code of practice suggesting a limit of six pickets at industrial disputes. While this is simply guidance and only applies to industrial disputes, the code is increasingly being adopted by the police at other types of picketing.

Trespass

A person is trespassing if present on private land without the permission, express or implied, of the landowner. A person can also be trespassing when operating outside the bounds of the authority given. For example, a right of way across a piece of land may give someone permission to pass across the land, but it does not give permission for other activities, such as protesting on the land.

Since local authorities own the surface of main roads and adjacent landowners often own other roads, these owners are able to take action for trespass against anyone who uses a road for anything other than passage.

While trespass itself is not a criminal offence, a landowner can use reasonable force to eject a trespasser who refuses to leave the land. In addition, the landowner can then apply to the court for an injunction stopping continuing trespass. It is then a criminal offence to break the injunction. Injunctions were granted by the courts during the miners' strike of 1984 to clear colliery entrances.

Sections 61 and 62 of the Criminal Justice and Public Order Act 1994 have strengthened the powers of the police to remove trespassers from private land in certain circumstances. For an offence to be committed under these sections there must be two or more trespassers, the trespassers must reside on the land (or intend to do so) and the landowner must have asked the trespassers to leave. In addition, before the police can require the trespassers to leave, they must consider either that one or more of them has damaged property on the land, used threatening, abusive or insulting language towards the landowner, or his/her family or employees, or that they have more than six vehicles on the land.

The provisions do not apply to trespassers living in buildings on the land. The courts have now also decided that to reside means to live somewhere regularly (although not necessarily each and every night).

Unfortunately, the term 'damage to land' is vague and may be open to wide interpretation. If this is the case, acts such as causing ruts in a path or chopping firewood could constitute causing damage. The interpretation also specifically includes the deposit of any substance capable of polluting the land, so if any part of the land is used as a lavatory this would seem to constitute damage.

Once the trespassers have been asked by the police to leave the land, it is an offence not to do so as soon as practicable. This means in less than a day in all but the most exceptional cases. Any trespasser failing to leave the land or returning within three months is then liable to be arrested. On conviction in the magistrates' court, the offence carries a maximum of three months' imprisonment and a fine of up to £2,500. The police have also been given the power to seize vehicles that are not removed.

Trespassory assembly

More recently, sections 70–71 of the Criminal Justice and Public Order Act 1994 (which retrospectively added to the Public Order Act 1986 sections 14(A–C)) introduced new offences where a demonstration also involved trespass (so-called 'trespassory assembly').

These powers were drafted primarily to deal with raves and to stop the annual solstice gathering at Stonehenge. However, the provisions were drafted in a much more general way.

Banning a trespassory assembly

Section 14 (A) allows the police to ban, in advance, an open-air assembly if it is to be held without the landowner's consent and if the assembly is likely to cause a serious disruption of the life of the community or significant damage to a site of historical, architectural, archaeological or scientific interest. The ban can be for up to four days and can cover a maximum circular area of a five-mile radius from a particular point.

The provisions only apply to anticipated trespassory assemblies and not existing assemblies, which cannot be banned. It is important to remember, however, that existing trespassory assemblies can be covered by other provisions such as breach of the peace.

A term such as 'serious disruption of the life of the community' is open to wide interpretation. Such disruption is likely to include traffic problems, noise problems, problems from inadequate sanitary facilities and damage to neighbouring property. In practice, this gives the police a wide discretion to ban these demonstrations.

As with the banning of marches, the police currently apply for the permission of the Home Secretary to impose a ban in London. Elsewhere, the police must apply to the relevant local authority who may impose a ban, with the consent of the Home Secretary.

The courts have recently confirmed that a person can be guilty of trespassory assembly even where the assembly is on the public highway and did not create an obstruction or public nuisance (see Chapter 14 for more on nuisance).

It is an offence either to organise or to participate in a trespassory assembly and the police can arrest those they suspect to be committing either offence. On conviction in the magistrates' court, these offences carry a maximum penalty of three months' imprisonment or a fine of £2,500 or both.

Preventing protesters travelling to a trespassory assembly

Section 14 (C) of the Public Order Act 1986 gives the police the power to stop people who they have reason to believe are travelling to a trespassory assembly and to direct them not to proceed. However, the police can only use these powers within the area specified in the banning order.

The police have the power to arrest anyone ignoring their directions and so committing an offence under the Act. The offence is tried in a magistrates' court. On conviction the maximum penalty is a fine of up to £1,000.

Aggravated trespass

A new offence of aggravated trespass was introduced by sections 68 and 69 of the Criminal Justice and Public Order Act 1994. It applies to trespassers who try to disrupt a lawful activity or to obstruct or intimidate those engaging in a lawful activity. A lawful activity is defined as any activity that can be done without either committing an offence or trespassing on land. It is no defence for protesters to argue that they disrupted a particular activity because they found it morally objectionable.

These provisions were brought in primarily to deal with hunt saboteurs, but have also been used against road protesters. They could also apply to other areas such as demonstrations by consumers outside a shop on property owned by the shop.

The offences only apply in the open air. As with trespassory assembly offences, a protester may have some right to be on the land (for example, exercise a right of way along a footpath or bridleway) but may commit an offence if this right is exceeded.

Generally, land forming the highway is not covered by these provisions and so protest on the highway will not be an offence under these provisions (although it may be an offence under other provisions such as obstruction of the highway).

It is an offence for a person to disobey the orders of the police to leave the land, if the police reasonably believe that the person has committed, is committing or is about to commit the offence of aggravated trespass.

The police may arrest a person if they have reason to believe that he/she is committing an offence under these provisions. The offence is tried in a magistrates' court. On conviction, the maximum penalty is three months' imprisonment or a fine of up to £2,500 or both.

These provisions have been used by the police to arrest protesters, so removing them from the protest, and then to release them without charge at the end of the protest.

Squats

Sections 72–76 of the Criminal Justice and Public Order Act 1994 contain enhanced provisions to allow the original owner to remove squatters living in buildings on land.

The owner can now apply to the magistrates' court for an interim possession order. The owner must tell the squatter in advance that he/she is applying for this order to allow the squatters time to file a sworn statement stating why they believe they have a right to occupy the building.

If an interim possession order is granted, it is an offence punishable by up to six months' imprisonment or a fine of £5,000 not to leave the premises within 24 hours of the notice being served on the squatter.

Camp and traveller sites

Sections 77–79 of the Criminal Justice and Public Order Act 1994
give a local authority the power to serve an order on those
camping on unauthorised sites, requiring that they move on. It is
an offence not to comply with the notice. The offence is tried in
the magistrates' court and on conviction carries a maximum
penalty of £1,000.

If the campers do not move on, the local authority can then
apply to the magistrates' court for an order allowing the local
authority to remove the vehicles, property and campers
themselves. It is then an offence to obstruct anyone involved in
the removal, and this offence is tried in the magistrates' court,
and on conviction carries a maximum penalty of £1,000.

LOCAL AUTHORITY BYLAWS

In additional to national law, the bylaws of individual local
authorities will often place additional restrictions on public
protest. Bylaws will usually cover activities such as the display of
posters, the use of loudspeakers, the distribution of leaflets and the
holding of meetings in certain places (or the need for consent for
such meetings).

The validity of these bylaws can be challenged on the
grounds that they are unreasonable. However, the courts have
been reluctant to question the validity of all but the most
unreasonable bylaw.

Breach of a bylaw is punishable by a fine not exceeding £500,
but there is also the possibility of a bind-over order. Local
authorities can also apply to the court for an injunction preventing
persistent breaches of bylaws. If an injunction is granted, any
further breach constitutes contempt of court.

In addition, the permission of the Department of the Environment, Transport and the Regions is needed before a person is
allowed to speak at certain places, such as Speaker's Corner in
Hyde Park or in Trafalgar Square.

OTHER PUBLIC ORDER OFFENCES

Public and private nuisance

A public nuisance is an act or omission that obstructs or causes inconvenience or damage to the public. This is a criminal offence and an injunction can be sought against it. There is also an associated civil action (that is, a court action between individuals) where the affected individual can seek damages for the loss suffered from the person responsible for the nuisance.

In order for something to be a public nuisance, it must affect the neighbourhood rather than one or a few individuals, in which case it would be a private nuisance. A private nuisance is not a criminal offence but the affected individual can still bring a civil action against the person allegedly committing the nuisance.

People have been found guilty of causing a public nuisance for a wide range of activities, including those involved in bomb hoaxes and those organising acid house parties. In addition, in the case of *Hubbard* v. *Pitt* (1976), the courts banned a group of social workers from distributing leaflets outside an estate agent's shop front, as the leaflets condemned the estate agent's practice of winkling out working-class residents from an area and selling the properties to 'yuppies'. The estate agent successfully claimed that the protesters were causing a public nuisance (and distributing libellous material) and was granted an injunction.

Often activities that would give rise to a public nuisance are controlled by local authority bylaws (such as noise control).

Police powers in London

There are obscure provisions dating from the nineteenth century that give additional powers to the police in London to control public order.

Section 52 of the Metropolitan Police Act 1839 and section 22 of the London Police Act 1839 give the Police Commissioner the power to make orders and give directions to prevent obstruction and to keep order. These powers can, therefore, be used to prevent assemblies or processions that are capable of causing an obstruction or likely to lead to a breach of the peace. These powers were used during the News International dispute at Wapping and also as the power underpinning anti-terrorist roadblocks in the City of London when they were first introduced in the early 1990s.

Local authority powers

Similarly, obscure provisions also give local authorities outside London powers to control public order. Specifically, section 21 of the Town Police Clauses Act 1847 gives local authorities the power to make orders in relation to the route that vehicles should follow to prevent obstructions in certain areas outside London.

Demonstrations near the Houses of Parliament

Demonstrations near the Palace of Westminster are controlled by orders made by the House of Commons at the beginning of each parliamentary session. These orders require the police to disperse assemblies or processions if they are likely to cause an obstruction, disorder or annoyance. These provisions only operate when Parliament is sitting.

Where somebody obstructs the police while they are enforcing these orders, the person can be arrested and charged with obstructing the police. These powers were used in the 1980s against OutRage, the gay campaigning group, when they protested against anti-gay legislation during the opening of Parliament.

Tampering with goods

Under section 38 of the Public Order Act 1986, it is an offence to tamper with goods or to claim to do so as a means of publicising a cause. Tampering with goods includes adding harmless substances (such as food dye) or placing a sticker on a product.

This charge is tried in either the magistrates' court or the

Crown Court. On conviction in the Crown Court, the maximum penalty is ten years' imprisonment or an unlimited fine or both. On conviction in the magistrates' court, the maximum penalty is six months' imprisonment or a fine of up to £5,000 or both.

TAKING ACTION AGAINST THE POLICE

As has been seen, the police have wide powers to take action in relation to public order matters. Nonetheless, there is a general principle that a police officer can only interfere with a person's liberty if he/she has legal authority to justify these actions. So if, for example, a protester is arrested for an offence for which he/she cannot be arrested, the police use excessive force in making an arrest, or a protester is searched without lawful authority, then there are a number of actions the aggrieved person can take.

Self-defence

If a person is unlawfully restrained he can use reasonable force to escape. However if a person uses excessive force to accomplish this, it could constitute assault. Unfortunately, since the police have very wide powers to detain and arrest protesters, it is not advisable, except in the most clear-cut of cases, to use any form of violence against the police.

A complaint against the police

Complaints against the police are conducted and investigated by the police themselves. Any complaint must be brought within a year of the incident being complained about and the complaint must be made against a particular officer or officers and must not be a general complaint about police policy.

If a person thinks that he/she may lodge a complaint or take legal action then it is crucial that the person gathers as much information relating to the incident as possible. This information should include, where possible, the name and police number of the officer concerned and the number of any police vehicle involved, as well as the names and addresses of any witnesses.

While the complaint can be made verbally it is normally better to submit it in writing and to keep a copy. Therefore, the person should not feel pressurised into making a verbal complaint at a police station. It is also important to remember that police officers falsely accused of wrongful conduct can sue their complainants for defamation.

There are two different possible procedures depending on the seriousness of the complaint.

Informal resolution

Minor complaints (such as rudeness to a member of the public) can be dealt with by a system of informal resolution. The chief constable of the police authority concerned appoints an officer (above the rank of the officer being investigated) to investigate the complaint. Following an investigation, there may be an explanation of the actions or an apology.

The complainant must consent to this system of informal resolution being used. If the complainant does not consent or is unhappy with the outcome, or if the chief constable considers that, having investigated the complaint, it is more serious than at first thought, then the matter must be referred for a formal investigation.

Formal investigation

More serious complaints (such as alleged assaults) are subject to a formal investigation. In these cases, the police investigation is overseen by the Police Complaints Authority (PCA). The PCA consists of up to 12 members, none of whom can be or can have been a police officer. The chair is appointed by the Queen and the others appointed by the Home Secretary.

Unfortunately, the PCA has resources sufficient only to supervise a small number of the total police investigations. While they must supervise complaints relating to death or serious injury or serious assault, they are not obliged to, and often do not, supervise 'less serious' complaints.

First of all, the chief constable of the police authority of the police officer subject to the complaint appoints an investigating office from the same or a different police authority to investigate the complaint. The PCA can influence the decision of whom to appoint. The investigating officer then interviews and takes statements from both the police officer who is the subject of the complaint and the complainant. Following this, and any other steps that the investigating officer considers necessary, he/she writes a report to the chief constable. In his/her report, the investigating officer will state whether he/she considers that the papers should be sent to the Director of Public Prosecutions (DPP), who will consider whether a prosecution is appropriate, and/or whether he/she considers that disciplinary charges should be brought. The PCA also receives a copy of the report (irrespective of whether the authority supervised the investigation of the complaint or not) and has the power to require that the papers be sent to the DPP and/or that disciplinary charges are brought.

If disciplinary charges are brought, the complainant is asked to give evidence at a disciplinary hearing tribunal. This tribunal consists of the chief constable of the police authority subject to the complaint and two members of the PCA. The complainant is allowed to bring a friend, but not a lawyer, to the tribunal hearing. In the course of the hearing the complainant is likely to be able to question the officer. The tribunal as a whole decides the guilt or otherwise of the police officer, but, if found guilty, the chief constable alone decides on the nature of the punishment.

The formal complaints procedure results in very few prosecutions or disciplinary charges being brought against police officers.

Civil action

A person is able to take legal action to seek damages in order to compensate him/her for the injuries received or the losses suffered as a result of a police action. For example, an action can be brought for false imprisonment, trespass to the person or trespass to goods, and assault and battery. Similarly, an action for malicious prosecution or for bringing a prosecution without good reason can be brought.

In order to pursue a civil claim a person will normally need to consult a lawyer, and legal aid may be available if the person is financially eligible (see Chapter 3). A civil claim has the advantage that a specific officer does not have to be named and the action can be brought against the chief constable of the particular force.

Private prosecution

If a police officer commits a criminal offence then a person can bring a private prosecution against the police. Usually, the person should make a complaint, wait for the matter to be investigated by the police and wait to see if the matter is referred to the DPP and whether the DPP bring a prosecution. If the police officer is convicted he/she may be ordered to pay the complainant compensation.

If the DPP fails to bring a prosecution, the complainant can bring his/her own prosecution. However, if a private prosecution is being considered, the person should obtain legal advice. As no legal aid is available for these types of cases, they can be very expensive.

Complaints against police policy

If the complaint concerns a more general criticism about the policing of a protest rather than a specific incident or the behaviour of a specific police officer, then the complaint can be made to either the Home Secretary or the Police Consultative Committee.

Independent inquiry

The Home Secretary is able to set up an independent inquiry into any police matter. However, in all but the most serious of cases it is unlikely that the Home Secretary will agree to an inquiry.

Complaints to the Police Consultative Committee

A person can also make a complaint to the Police Consultative Committee regarding policing policy in a particular area. These committees exist in nearly all the police authority areas but have only a consultative role and so have no power to actually regulate police policy.

CONCLUSION

For any group contemplating a form of public protest it can be seen from this chapter that the police hold wide powers to detain and arrest demonstrators for a whole variety of offences. This should not discourage the group, because the reality is usually very different, and the police are not generally keen on being seen to be too heavy-handed (although there are numerous exceptions to this). Further, with growing public awareness and sympathy, this form of protest is clearly an increasingly effective way of bringing public attention to an issue.

Whether public demonstrations are successful depends very much on the circumstances. In many situations the developers, the government and the police dig their heels in, unwilling to be seen to be backing down in the face of public protest. There are cases of successful demonstrations, however, and, at the very least, taking these steps ensures that environmental concerns remain centre stage within our society. It can be a case of losing the battle but winning the war.

11 Pollution Control – The Statutory Framework

The Ancient Greeks were concerned about wells being poisoned by contamination and Plato suggested the death penalty for such contamination in his 'Laws'. Those polluting the London air by burning coal were threatened with 'grievous ransoms' by a 1306 proclamation. Forty years later a fine of two shillings could be levied on householders who failed to remove filth and refuse from outside their dwellings.

Developments in legislation were piecemeal until the onset of the industrial revolution provoked the need for greater pollution control. Halting progress was made in the nineteenth century, although, as today, vested industrial interests effectively watered down much of the legislation. The concept of Best Practicable Means for pollution control was introduced as a secure defence for polluters from 1842. This concept was to endure until the early 1990s and is still used in Northern Ireland. However an important step forward was the formation of the Alkali Inspectorate, set up primarily to deal with the 'monster nuisance' of alkali plants producing sodium carbonate for the manufacture of soap, glass and textiles – and discharging into the air hundreds of tonnes of hydrochloric acid by-product. The Alkali Act 1863 gave the inspectors powers to enter factories, not to protect workers but on behalf of the inanimate air and environment – an important step in the development of environmental law. At one point the Inspectors salary increases were blocked on the basis that they had not effectively stopped the damage to local vegetation – an incentive that it may have been interesting to keep!

At this time, public health legislation introduced arrangements to provide for improved water supplies and sewerage systems, necessitated by the growth of larger towns and cities. However, a truly coherent approach to pollution control has only really developed over the last 30 years. Indeed, until very recently the focus of practically all control remained on public health, rather than on environmental protection *per se*. Many aspects of current pollution control regimes still reflect the piecemeal public-health-based origins.

Although the primary focus of this book is on private law remedies available to deal with existing pollution, it is useful also to be aware of public law mechanisms that can be used to prevent or limit pollution before it happens, as well as those intended to deal with existing pollution. The statutory framework includes major strands with the potential to achieve both these objectives.

The regulators

In order to be able to intervene to greatest effect, campaigners need to be aware which public bodies have environmental protection duties, what those duties are and how they can be enforced. Most prominent is the Environment Agency, which was created on 1 April 1996 with the merging of the National Rivers Authority, Her Majesty's Inspectorate of Pollution and Waste Regulation Authorities. The pollution control duties of the Agency include major industrial processes under integrated pollution control (IPC), waste management and licensing, and regulation and control over most water pollution. Local authorities, primarily district or unitary councils, are responsible for most of the remaining functions, including controls over air pollution from smaller processes, statutory nuisance and measures for dealing with contaminated land. Discharges of industrial wastes to sewers are regulated by the privatised water companies – the only example, so far, of a privatised environmental regulator in the UK.

There are some encouraging developments in the philosophy of pollution control coming from the United States and Europe following critical assessment of the failures of the current systems.

These include moves towards 'zero discharge' of toxic and persistent chemicals and a reversal of the onus of proof so that companies have to demonstrate safety before they can release chemicals. Unfortunately UK authorities are often lagging some way behind the leading edge in this debate, and UK delegations at international conferences have generally argued against such measures. Instead the philosophy here has been based on 'prove harm', where companies are basically allowed to discharge what they like unless it is proved to be harmful, in which case the consent would be revised. This approach is not precautionary and it may take many years before the impacts of a mixture of chemicals emitted from a site are shown to be damaging to the environment or to the health of local people. There is a lot to be done to improve this situation and the law can be a useful tool.

The main statutory provisions are now consolidated in the Environmental Protection Act 1990 (EPA 1990), the Water Resources Act 1991 and the Environment Act 1995. It is only possible in this chapter to provide a brief overview of the main areas, as a detailed survey of all of the relevant provisions would fill a book on its own. There are a number of ways in which campaigners can intervene effectively in these statutory processes.

Applications for consents

Most industrial-scale polluters must apply to a regulatory authority for consent before they can embark on a polluting activity. Although it is more difficult to feel involved in the pollution control process than with planning applications, this process provides an opportunity for concerned citizens to highlight the potential problems with the consents that are sought. It is important to find out about the proposals (see Chapter 5 for more on obtaining information) and to ensure that concerns are brought to the attention of the regulator. Pressure can also be brought to bear through the media, local councillors and MPs. The input of campaigners has the very important function of changing the focus of what can be a very cosy relationship between licensing bodies and polluters. Where environmental concerns and/or procedural requirements are not met, a judicial review or a statutory appeal might well be a way of forcing the regulator to take more notice of what campaigners are saying.

Enforcement

Public bodies charged with licensing or authorising functions also always have monitoring and enforcing functions. Unfortunately, these crucial aspects of their responsibilities are not always carried out with the degree of rigour and enthusiasm one might expect. Citizens can play an important part in highlighting breaches of pollution control using the media. High-profile coverage of the 472 breaches of discharge consents at the ICI Runcorn site between 1995 and 1997, for example, has been instrumental in getting the Environment Agency to crack down on the operation of that plant. An alternative legal option, in some circumstances, can be a challenge in the courts of the failure of regulators to act. This is done by way of judicial review (see Chapter 9). Campaigners could also consider taking action directly against polluters. This can be done by means of civil actions, or it may be possible to bring a private criminal prosecution against a polluter (see below in this chapter). Such actions help, not only by challenging the illegal actions of polluters, but also by highlighting the inaction of an ineffective regulator.

AIR POLLUTION

Ever since the discovery of fire more than half a million years ago, air pollution has been a problem. Historically, concerns fall into three main categories. The first includes the effects of smoke produced by the burning of coal for domestic and industrial purposes. Smoke and clean air is now regulated under the Clean Air Act 1993. The second category relates to toxic emissions, initially from alkali works, that impact directly on public health and the environment. This is now mainly dealt with under Part I of the EPA 1990, under IPC for the more polluting processes and under local authority air pollution control for the smaller works. The third, most recent category of concerns relates to pollution that impacts further away from its source. These problems of global warning, ozone depletion, bio-accumulation of toxic emissions and transnational acid rain are particularly difficult to deal with as they really need coordinated international responses. After the United Nations Conference on the Human Environment in 1972, the first serious steps to deal with international air pollution problems were taken, particularly with treaties on acid rain, and European Community action on polluting factories and, in the late 1980s, ozone-depleting substances.

Clean Air Act 1993

District councils and unitary authorities are charged with respon-sibilities under this Act for controlling visible air pollution such as soot, ash and grit. Under the Act it is an offence for dark smoke to be emitted from a chimney or from industrial or trade premises. Breach of these provisions, subject to certain defences, is a criminal offence and the occupier of the premises and/or the person who causes or permits the emissions is liable. These regulations also apply to emissions from domestic chimneys.

Emissions of grit and dust from industrial furnaces are regulated by the Act, as is the requirement to fit grit- and dust-arresting plant on new furnaces. Provisions allowing local authorities to declare smoke-control areas constituted a major means of regulating pollution – primarily domestic pollution – from coal. With the growth in use of gas and oil and the use of smokeless solid fuel, these provisions are of less importance in recent years. The 1993 Act provides that the controls under the Clean Air Act do not apply to processes that are regulated under either IPC or local authority air pollution control.

Local authority air pollution control

For many industrial processes, the scale of the plant determines whether it is subject to local authority air pollution control (LAAPC), known as 'Part B', and regulated for emissions to air alone, or is regulated by the Environment Agency as part of IPC, known as 'Part A'. Lists of processes that are LAAPC processes are found in the Environmental Protection (Prescribed Processes and Substances) Regulations 1991 (as amended). For example, a clinical waste incinerator that can burn less than one tonne of waste per hour would be regulated by the local authority, while a larger unit would be covered by IPC and regulated by the Environment Agency.

For Part B processes, an application must be made to the appropriate local authority giving details of the location of the process, the name and address of the operator and a description of the process involved. It must also give details of prescribed substances to be released from the plant, with an assessment of the environmental consequences of the emissions and the proposed methods of abatement and monitoring. It is a requirement that issues relating to BATNEEC (the 'best available techniques not entailing excessive cost') be addressed. Upon receipt of an application, advertising and public consultation is undertaken and representations are invited within 21 days of the publication of the details of the application. Applications for new processes must be determined within four months, with a period of up to nine months allowed for existing processes.

The local authority will either grant the application subject to conditions or refuse the application. If the authority considers that the applicant will not be able to comply with conditions, it is required to refuse to authorise the process. Within each authorisation there is an express or implied condition that BATNEEC will be used to prevent the release of prescribed substances into the air or otherwise to reduce the release to a minimum or to render it harmless.

The Department of the Environment, Transport and the Regions produces process guidance notes for all Part B processes. These contain recommendations on emission limits (which are generally less stringent than those required under IPC), and requirements for record keeping, reporting and monitoring. The requirements set out in the Department guidance are authoritative, but authorities can depart from them if they give good reasons. Applicants aggrieved by refusal or imposition of conditions have a right of appeal to the Secretary of State for the Environment, Transport and the Regions (referred to throughout this chapter simply as the Secretary of State). These appeals, which are quite rare, will be determined either by way of written representations or by a hearing (see Chapter 9 for further information about how to deal with appeals).

Local authorities have a range of powers to enforce their air pollution controls and environmental health officers have wide powers to investigate and inspect plants falling within the scope of the Act. Failure to comply with the provisions of the Act is a criminal offence.

For citizens dissatisfied with the approach taken by the local authority, a number of possible remedies should be considered. There is the possibility of taking a private prosecution of the polluter for breach of the requirements of Part I of the EPA 1990 (see below). Alternatively, there may be scope for a judicial review of the failure of the local authority to carry out its statutory duties under Part I or Part III of the EPA. Another route might be to take action under section 82 of the EPA 1990 against the polluter for statutory nuisance (see Chapter 14, pages 358–63). Information relating to processes regulated under LAAPC, including their emission data, must be placed on the public register, but in practice the quality of these registers varies greatly between authorities.

International issues

The issues relating to international and global air pollution problems are very complex and largely outside the scope of this book. However, it is appropriate to try to summarise briefly the framework that is in place. In order to deal with these problems, the need for a shift from the use of reactive legal controls to a more proactive policy-based approach has been recognised. Britain has been a past master at exporting pollution and our stance on many of the developments has not been progressive. The first steps have been embodied in the Environment Act 1995. Increased air pollution monitoring is now being introduced.

European Union air pollution measures

European directives have set quality standards for sulphur dioxide
(80/779/EEC), lead (82/884/EEC) and oxides of nitrogen (85/203/
EEC). These have been implemented in the UK by the Air
Quality Standards Regulations 1989, which impose an obligation
upon the Secretary of State to ensure that levels of sulphur
dioxide, nitrogen dioxide, lead and smoke do not rise above EU
limits and to set up a system of sampling stations to measure air
quality. Emission limits are set by a Framework Directive
(84/360/EEC), with more specific controls on emissions from large
combustion plants (Directive 88/609/EEC) and new municipal
waste incinerators plants (Directive 88/369/EEC). Vehicle emissions (Directives 88/76/EEC, 89/458/EEC and 91/441/EEC), lead in
petrol (Directive 85/210/EEC) and sulphur in gas oil (Directive
87/219/EEC) have also been controlled. (See Chapter 6 for more on
the implementation of European directives.)

Vehicle emissions

It is increasingly recognised that the massive growth in numbers of
motor vehicles and the resulting emissions have made a substantial
contribution to air pollution problems. Controls on vehicle emissions have been focused in two main ways. By section 30 of the Clean
Air Act 1993, the Secretary of State has powers to make regulations
imposing controls on fuel content, which have been implemented to
bring into effect the requirements of the relevant European directives referred to above. The second way is by means of control on
vehicle design. The Road Vehicles (Construction and Use) Regulations 1986 have been regularly amended to take account of requirements of European directives providing minimum standards for
emissions of carbon monoxide, hydrocarbons and oxides of nitrogen. All new vehicles must be capable of running on lead-free fuel.

An important source report for campaigners on air pollution is
The United Kingdom National Air Quality Strategy, which was published in March 1997 and sets provisional targets for a range of pollutants. When launched it was criticised by the Labour Party, then

in opposition. The new Labour Government announced in July 1997 that it would try to bring forward the implementation dates from 2005.

WATER POLLUTION

The consent system

A consent is required from the Environment Agency for the discharge of any trade or sewage effluents into 'controlled waters'. These include virtually all inland and coastal waters, such as rivers, canals, lakes, reservoirs, ground water and areas of sea near the coast. The system for acquiring consents is set out in Schedule 10 of the Water Resources Act 1991 and involves some degree of public involvement. Applicants apply to the Agency, which has a discretion as to the details required. The Agency must generally publicise the application in a local newspaper and the *London Gazette*, and notify any relevant local authorities and water undertakers at the applicant's expense. The Agency must take into account written representations made within six weeks of publication of the application, and it has power to grant consent, either unconditionally or subject to conditions, or to refuse consent.

A consent is required from the Environment Agency for:

- any discharge of trade or sewage effluent into 'controlled waters';

- any discharge of trade or sewage effluent through a pipe from land into the sea outside the limits of 'controlled waters';

- any discharge where a prohibition is in force.

A prohibition is a means by which discharges can be prevented on a selective basis, where a blanket ban is not justified.

Water-quality objectives

The Water Act 1989 introduced statutory water-quality classifications and objectives for the first time. The provisions were reproduced in the Water Resources Act 1991, sections 82–84. They promised a great deal but have yet to deliver, having proceeded at a snail's pace through an arduous series of consultations. Ten years after the objectives were first proposed, the scheme is set to be tried out on only eight rivers. Assuming that these trials proceed without further delays, there is now real potential for enhanced protection of rivers.

There is a three-stage process for the setting and maintaining of statutory water-quality objectives. The first stage is the setting of classification systems for waters under section 82. Four sets of regulations were set up to do this: (1) the Surface Waters (Classifications) Regulations 1989, (2) the Bathing Waters (Classification) Regulations 1991, (3) the Surface Waters (Dangerous Substances) (Classifications) Regulations 1989 and 1992 and (4) the Surface Waters (River Ecosystem) (Classification) Regulations 1994. These regulations classify waters into categories depending on their quality, and on levels of certain substances that should not be exceeded.

The second stage is that water-quality objectives for individual stretches of controlled waters may be set by the Secretary of State. The choice of an objective by the Secretary of State would mean that the appropriate standards laid down for that objective in the classification regulations would apply to that stretch of water. The procedures for setting statutory water-quality objectives involve at least three months' publicity of the proposed objective and all representations and objections should be considered. A public local inquiry may be held.

The third process is that under section 84 the Environment Agency and Secretary of State are placed under a duty to exercise their powers under the Water Resources Act 1991 so as to achieve statutory water-quality objectives at all times, so far as it is practicable to do so. Judicial review could potentially be used if the Agency fails to ensure enforcement of the duty (see Chapter 9).

Public registers

Under section 190 of the Water Resources Act 1991, a public register must be kept by the Environment Agency of, among other things, all applications for consent, consents granted, conditions attached to a consent and the results of any samples of receiving waters or of effluent. The register must be open for inspection by any member of the public free of charge during normal office hours, and reasonable copying facilities must be available at a reasonable fee. The register is an important database for campaigners wishing to monitor water quality. It can be used to mount a private prosecution, or to provide evidence of a civil claim, or to provide general information on the state of the water environment.

Water pollution offences

There are offences relating to breaches of the above consents and also a general offence of polluting controlled waters. Under section 85 of the Water Resources Act 1991 it is an offence for a person to:

1 cause or knowingly permit any poisonous, noxious or polluting matter or any solid waste matter to enter any controlled waters;

2 cause or knowingly permit any trade or sewage effluent to be discharged into controlled waters; or

3 contravene the conditions of a discharge consent granted as part of an IPC authorisation, or otherwise.

It is a defence to the first two of these offences to show that the discharge was in accordance with a discharge consent, as discussed above. However, it is an offence to breach any conditions attached to a consent. It seems to be accepted that all three are 'strict liability' offences – that is, there is no need to show intention or negligence on the part of the defendant. For the purpose of offences 1 and 2 above, it would seem sufficient for the prosecutor to show that there had been a discharge into controlled waters and that the discharge emanated from the defendant's land or operations.

The offence of 'causing' a pollution offence requires a positive act on the part of the defendant. However, in the case of the construction of a drainage system (which later overflowed) it was held that controlling the industrial activity that used the drainage system to the point of overflow was sufficient to 'cause' the offence. The industrial operator rather than the developer who built the system was the true defendant.

The offence of 'knowingly permitting' is more limited, in that there is the requirement of knowledge, but this may be used in situations where a person is passive even after knowing of an incident. 'Permitting' is the act of allowing something to occur, or, more specifically, not preventing it, if it was possible to do so. 'Knowingly' has been interpreted to mean 'with actual knowledge', or 'refraining from enquiry for fear of the truth'.

Under section 118 of the Water Industry Act 1991 it is an offence for an occupier of trade premises to discharge trade effluent into public sewers in breach of a sewerage undertaker's consents as to the nature and composition, quantity and rate of the discharge and the identification of the sewers into which the discharge can be made.

The penalties

These are 'offences triable either way'. In the magistrates' court there is a maximum prison term of six months or a maximum fine of £20,000 for each offence committed. In the Crown Court there is imprisonment of between two and five years, depending on the offence, and an unlimited fine.

Company directors, officers and anyone acting in such a capacity may be jointly liable for the criminal activities of the companies they run, and anyone whose act or default caused the offence can also be prosecuted for it. Thus employees and others involved in running the company may be caught.

Clean-up costs

Under section 161 of the Water Resources Act 1991, as amended by the Environment Act 1995, the Environment Agency has the power to serve a 'works notice' in respect of controlled waters. This notice can be served on anyone who caused or knowingly permitted polluting matter to enter water, or on directors of the company involved or on any person whose act or default caused the offence.

The notice requires the person carrying out the specified work to clean up the water and it is an offence for the person served to fail to comply with any requirements of the notice. The Environment Agency can also carry out the works itself in certain circumstances and recover any costs reasonably incurred in doing so. It can carry out investigations to determine who was responsible for the pollution and also recover those costs.

Although no criminal prosecution needs to be brought in these cases, the Agency has a tight budget and given the practical difficulties of recovering costs it is likely to use these powers very sparingly.

Civil liability

In some ways civil liability represents a greater threat to polluting companies than criminal liability because of the potentially large damages that can be imposed and the commercial impact of prohibitory injunctions. A civil action can be brought even though no criminal offence has been committed, and as the standard of proof is lower it stands a much greater chance of success.

Civil liability generally will arise at common law and the areas most likely to be relevant are nuisance, negligence and the rule in *Rylands* v. *Fletcher* discussed in Chapter 14 (pages 356–8). There may also be a statutory nuisance; section 259(1)(a) of the Public Health Act 1936 provides that any pool, pond, ditch, gutter or watercourse in a state that is prejudicial to health or a nuisance is a statutory nuisance. If the pollution results from an industrial process then it may be regulated by the IPC regime (see below, page 271). (See also Chapter 15 for more on water cases.)

European Union water pollution measures

European directives have been extremely important in determining the course of water protection legislation in the UK. This has not been without tension – Britain has always argued for a standards to be based on receiving waters rather than discharge levels.

Among the most significant directives are the framework directive on Dangerous Substances in Water (76/464/EEC), the Bathing Water Directive (76/160/EEC) – for which compliance should have been achieved by 1985, but which won't be achieved for some beaches until the next century – and Directive 80/778/EEC, setting standards for water for human consumption.

WASTE MANAGEMENT

Every nine months we produce more than enough waste in the UK to fill Lake Windermere. Most is agricultural, mining or construction waste, with only one twentieth coming from our households. Nevertheless, household waste still amounts to over 20 million tonnes per year, with commerce and industry generating over four times this amount (85 million tonnes).

One traditional method of waste disposal is landfilling holes in the ground. The many problems with this method have at last been recognised. When landfilled materials break down they may produce gases that escape into the air or pollute water supplies. Another traditional method is discharge to sewers, to rivers and streams or to the sea. The resulting pollution impacts greatly on the environment and also on public health. More recent methods employed include dumping at sea and incineration, and these also have substantial associated difficulties. The consequences of waste disposal can be long-term environmental problems that are very difficult and costly to remedy. There are no easy answers. A new approach is therefore developing, which focuses on waste minimisation, reuse and recycling.

In order to develop strategic control of the process of waste disposal, various forward-looking methods have been introduced. Waste-disposal plans were drawn up by some Waste Regulation Authorities (WRAs), before they were merged into the Environment Agency. In England 20 authorities had plans or statements approved although eight were subject to amendment. These plans deal with the strategic approach to be taken to treatment and disposal of waste within the authority's area. They are complimented by waste local plans (which are incorporated into unitary development plans where they exist) produced as part of the development plan process undertaken by local planning authorities (see Chapter 7, page 112). These deal with the siting of and restrictions upon waste-management development within an area. Planning permission is generally required for storing, handling or disposing of waste. Applications are dealt with by county (or unitary) planning authorities.

The Environment Agency is currently engaged in research to develop a national waste strategy based on the 1995 white paper 'Making Waste Work'.

European Union waste control measures

A number of European directives dealing with waste have been adopted. The Framework Directive on waste 75/442/EEC, amended by Directive 91/156, establishes requirements for planning and licensing waste-disposal operations. Definitions of what constitutes waste are also set out in the Directive. Additional controls are also imposed on hazardous waste by Directives 78/319/ECC and 71/689/EEC, which also defines hazardous waste. Other directives dealing with sewage sludge (Directive 86/278/ EEC), transfrontier shipment of waste (Regulation 259/93/EEC) and incineration of waste (Directives 89/369/EEC and 89/429/ EEC) are also in place.

What is waste?

Determining what waste is has caused surprising difficulties. Waste is now defined by means of the Framework Directive as amended. In very general terms it is defined as 'any substance or object which the holder discards or intends or is required to discard'. This definition is dealt with in much greater detail in the Waste Management Licensing Regulation 1994 and is being clarified by a series of cases in the European Court of Justice.

Previous definitions of waste have now been amended to fall into line with the concept of 'directive waste'.

Waste-management licensing

The waste-management regime has three main elements. It is an offence to treat, keep or dispose of controlled waste in a manner likely to cause pollution or harm to human health. Further, a duty of care is imposed on all those who deal with controlled waste with criminal sanctions for failure to comply with the duty. Finally, a waste-management licensing system exists, which includes the transport of waste.

A waste-management licence is required to treat, keep or dispose of controlled waste. The licence is granted subject to conditions and only to an applicant who is considered to be 'a fit and proper person'. In order to qualify as a fit and proper person an applicant:

- must not have been convicted of a waste-related offence;

- must ensure the licensed activities are managed by a 'technically competent person';

- must satisfy the agency that he/she is able to make adequate financial provision to cover the liabilities of the licence.

Licences may be revoked by the agency if it considers that the licensee is no longer a fit and proper person or if the continuation of the licensed activities would cause environmental pollution or harm to human health. In cases of urgency, licences can be suspended. The agency will only agree to transfer a licence to another fit and proper person.

In order to avoid the problems caused by operators walking away from polluting sites, a site licence can only be surrendered once the agency is satisfied that it no longer presents a pollution or health risk (the criteria for this are given in 'Waste Management Paper 26A'). A certificate of completion is then issued.

Duty of care

Any person who imports, produces, carries, keeps, treats or disposes of controlled waste has a 'duty of care' to take all relevant reasonable measures to prevent the commission by another person of a waste-related offence. This requirement in Section 34 of the EPA 1990 ensures that any person in the waste chain must take an interest in where the waste comes from and where its going to and how it is handled by these other people. In particular, it is also a requirement to ensure that any person to whom the waste is transferred is an 'authorised person' and that a written description of the waste is provided with it. A formal transfer note must also be completed. Details of the requirements of the duty of care are set out in the Environmental Protection (Duty of Care) Regulations 1991 and a circular and code of practice produced by the Department of the Environment, Transport and the Regions.

Waste carriers

Under provisions introduced by the Control of Pollution (Amendment) Act 1989, any person transporting waste is required to be registered as a carrier. The Environment Agency handles licences, which last for three years but may be revoked if offences are committed under the 1989 Act.

Criminal offences

This regulatory regime is underpinned by a number of criminal sanctions. Under Section 33 of the EPA 1990 it is an offence to treat, keep or dispose of controlled waste in a manner likely to cause pollution or harm to human health. It is also an offence to fail to comply with a waste-management licence or to breach the statutory duty of care. Transporting controlled waste without a licence is also an offence.

Public registers are maintained of the information relating to waste-management licences (see Chapter 5, pages 84–5).

INTEGRATED POLLUTION CONTROL

Traditionally, control over emissions to land, air and water has been governed by three separate statutory regimes. However, it has been recognised that emissions to all three media are closely related and that they need to come within a unified framework. There has also been pressure from Europe to integrate controls – culminating in the Directive on Integrated Pollution Prevention and Control (96/61/EEC). Integrated pollution control (IPC), introduced by Part I of the EPA 1990, set up a limited framework for such control. The most polluting industrial processes, known as 'Part A' processes, are now regulated by the Environment Agency. The Agency must exercise its powers 'for the purpose of preventing or minimising pollution of the environment due to the release of substances into any environmental medium' (section 4(2), EPA 1990). As explained above, the remaining processes are controlled by local environmental health authorities for air pollution purposes only. In such cases, for example, discharges to water are regulated by the Environment Agency. The Secretary of State determines by way of regulations which processes fall into Parts A or B, generally on the basis of the scale of the operation. Listed substances, which are of particular concern, are also specifically controlled, including the 'red list' of dangerous substances.

Part A processes

The EPA 1990 requirement to have an authorisation in order to operate an IPC process was phased in by industrial sector over several years following the implementation of the Act. Operating a Part A process outside the conditions of, or without, a valid authorisation and the other requirements of the Act is an offence (section 23(1), EPA 1990).

Applications for authorisation, which are frequently very lengthy and technical, are submitted to the Environment Agency. Full details are set out in the Environmental Protection (Applications, Appeals and Registers) Regulations 1991. The information required includes a detailed description of the process and the quantity and nature of any prescribed substances to be released. Any techniques used to prevent or minimise the release of such substances must also be described, as must arrangements for discharge of waste produced. A description of arrangements to monitor release of substances is also required in the application. If necessary, the Agency can also require further details of any relevant matters.

Notification and consultation

The operator of a process must advertise in a local paper any application for authorisation or any variation that involves a substantial change (for example, increased emissions). The advertisement must appear between 14 and 42 days after the application or notice of variation is made. Various statutory bodies are also consulted.

Although rarely used, the Secretary of State has power to call in applications and, following a public inquiry or an informal hearing, to direct the Agency whether to grant the authorisation and, if so, subject to what conditions.

The general principle of freedom of access to information regarding an application is qualified, in particular by the exception relating to commercial confidentiality. The applicant must seek approval from the Environment Agency to exclude any such information from the register. The applicant has a right of appeal to the Secretary of State against a refusal. The withholding of information on grounds of commercial confidentiality is a significant problem for environmental campaigners. It seems that the Agency is often willing to withhold information, almost automatically, on the suggestion by the applicant that it is commercially confidential (see Chapter 5, page 90, for more on this).

Determination of applications

Following receipt of representations, the Environment Agency may then refuse or grant the authorisation subject to conditions. In making this determination, the Agency must have regard to achieving the objectives set out in section 7(2) of the EPA 1990, including:

1 That in carrying out the process 'best available techniques not entailing excessive cost' (BATNEEC) will be used:
   (a) to prevent the release or if not practicable to reduce the release of prescribed substances to a minimum and to render harmless any such substances released
   (b) to render harmless any other substances which might cause harm if released.

2 Compliance with directions of the Secretary of State given to implement EU or other international legal obligations to protect the environment.

3 Compliance with limits and achievement of quality standards or objectives prescribed by the Secretary of State.

A condition is implied into every authorisation that the process will be carried on using BATNEEC. This is intended to ensure continuing pressure on polluters to upgrade their process using new available technology to minimise pollution.

IPC Guidance notes are used by the Environment Agency to provide information about standards achievable for every regulated process. BATNEEC is not defined in the Act but is dealt with in *Integrated Pollution Control: A Practical Guide*, produced by the Department of the Environment, Transport and the Regions, and regularly updated.

Appeals

Refusals of IPC applications are rare but applicants do have a right of appeal to the Secretary of State against refusal to grant or vary an authorisation or to impose conditions, or against any enforcement measures. Appeals are heard either by way of written representations or by means of a hearing, similar to planning appeals (see Chapter 7, page 136). All appeals on IPC are determined by the Secretary of State, who appoints an inspector to conduct the appeal and to report back with recommendations. The Secretary of State will then produce a decision.

Enforcement

The Environment Agency has a range of enforcement powers to ensure that authorisations are complied with, and failure to comply with the requirements of an authorisation or any enforcement action is a criminal offence. It is triable either in the magistrates' court (maximum imprisonment of six months or £20,000 fine) or in the Crown Court (maximum two years' imprisonment or an unlimited fine). In certain circumstances, directors and managers of a company can be prosecuted for their part in any offence.

In practice, however, informal persuasion rather than formal enforcement action is overwhelmingly employed by the Agency in the age-old tradition of pollution control in this country. As a consequence, prosecutions by the Environment Agency – and, previously, Her Majesty's Inspectorate of Pollution (HMIP) – have been very limited in number. Where campaigners are concerned that breaches of IPC are not being properly enforced, the possibility of private prosecution should be considered (see page 279 below).

CONTAMINATED LAND

Many of the measures already mentioned in this chapter are aimed at reducing the risk of contamination of land. There are, in addition, provisions for identifying and then dealing with land once it has been contaminated. Civil liability for land is dealt with to some extent by the common law. In addition, Section 73(6) of the EPA 1990 provides that a person who illegally deposits waste on land is liable for the damage it causes, not only in criminal law but also in civil law.

With perhaps 100,000 contaminated sites in the UK, creating a legal framework to deal with a clean-up of contaminated land has proved to be a particularly difficult problem. Proposals to introduce a register of land that may be contaminated were included in the EPA 1990 but dropped in 1993, in response to pressure from large landowners who feared that inclusion on the register would reduce the value of land.

Part IIA EPA 1990 – Contaminated Land
Provisions to deal with contaminated land clean-up were introduced by the Environment Act 1995, which inserted a new Part IIA (sections 78A–78YC) into the EPA 1990. Although the statutory provisions provide a framework, much of the most important material is contained in government guidance, which local authorities must have regard to (section 78B(2), EPA 90).

Definition of contaminated land
Contaminated land is defined as 'any land which appears to the local authority in whose area it is situated to be in such a condition, by reason of substances in, on or under land, that

(a) significant harm is being caused or there is a significant possibility of such harm being caused;
(b) pollution of controlled waters is being or is likely to be, caused.'

'Harm' is defined as 'harm to health of living organisms or other interference with the ecological systems of which they form part and, in the case of Man, includes harm to his property'.

These matters are dealt with in greater detail in guidance notes, which advise that harm should be disregarded for purposes of contaminated land unless the substances are likely to cause serious and lasting harm or damage to human health, ecosystems or property. This assessment of the possibility of harm is to be carried out with reference to the effects of the contamination and principles of risk assessment.

The new mechanism introduced in Part IIA of the EPA 1990 borrows the legal framework used in Part II of the Act dealing with statutory nuisance (see Chapter 14). Local authorities are required to identify contaminated sites and then decide what action needs to be taken by way of remediation. A remediation notice may then be served, and non-compliance is a criminal offence. The local authority can also carry out works in default and recover the reasonable costs involved.

Identification of contaminated sites

District or unitary authorities have responsibility for identification of contaminated sites under a duty to inspect land in their own area from time to time. In cases of water pollution, the Environment Agency must be consulted. Where a site may be particularly hazardous, the local authority must refer it to the Agency to determine whether to designate it as 'a special site'. In those cases remediation becomes a matter for the Agency.

Once the authority has identified a 'contaminated' site it must determine what steps should be taken by way of remediation. The Act provides that remediation involves a number of steps, commencing with the assessment of the condition of the land or impact on water and adjacent land; works to be carried out to minimise, remedy or mitigate' the effects; and inspection of the land or water following the carrying out of works.

Remediation notices

The local authority (or the agency in the case of 'special sites') then drafts and consults on who is 'the appropriate person' to carry out the works to be undertaken and what timescale is considered appropriate. In determining the works to be carried out, the authority must undertake a cost–benefit judgment, ensuring that the cost of clean-up is justified by the seriousness of the harm that would be suffered if it was not carried out.

Other than in cases where there is an imminent danger of serious harm or of pollution of controlled waters, remediation notice may not be served until notification of and consultation with the 'appropriate person' has been carried out and a period of three months has elapsed from the date of notification.

In certain circumstances, the authority may decide not to serve a remediation notice. These include circumstances where the cost of remediation would be unreasonable in the light of the benefits to be achieved, or where voluntary works are being or will be undertaken. The determination upon whom the notice should be served is a particularly difficult and important one. Responsibility must be allocated between joint 'appropriate persons'.

The 'appropriate person' is the person or any of the persons who caused or knowingly permitted the contaminated substances or any of them to be present in, on or under the land. A person can only be the appropriate person in respect of the substances or their natural chemical products which he/she actually caused or knowingly permitted to be present and caused the land to be contaminated. If an appropriate person cannot be found, the owner or occupier of the land is then the 'appropriate person'. Owners and occupiers can only be held liable to clean up contamination for which the original depositor cannot be found.

Appeals

A person in receipt of a remediation notice can appeal within 21 days. The appeal lies to the magistrates' court in respect of a notice served by a local authority and to the Secretary of State in respect of a 'special site' where the notice was served by the Environment Agency.

Non-compliance with the requirements of a remediation notice, without reasonable excuse, is an offence. On conviction a fine may be imposed of up to £5,000 and additionally up to £500 per day for continued non-compliance with the notice. In respect of industrial, trade or business premises, the maximum fines are £20,000 and £2,000 per day.

The most important power is that, in the event of non-compliance with a notice, the authority serving the notice can carry out the works itself and recover the cost of doing so from the person served with the notice. Unfortunately, historically, local authorities have been very reluctant to use these powers as they do not have budgets to find these costs of clean-up. This is particularly likely to be a problem in cases of contaminated land where the costs may stretch to very large sums.

Escapes of substances to other land

One of the main problems of contaminated land is that the contamination may spread to adjoining land in different ownership or occupation. Under the statutory provisions, the original polluter can be held responsible for the contamination that escapes to other land. Owners and occupiers of the original land, who can be held to be 'appropriate persons', are not liable to carry out remediation works in respect of substances that have migrated to other land. Innocent purchasers of land from which contamination has escaped may also avoid liability for pollution caused by their predecessors in title.

Public registers

Local authorities are required to maintain a public register of remediation notices, charging notices, appeals and convictions. Owners and occupiers and other appropriate persons may also place on the register any information they wish regarding action they have taken to comply with remediation notices. Because of the tough tests for designating land as contaminated, the registers are likely to be very limited in extent. (See Chapter 5 for further information regarding public registers.)

PRIVATE PROSECUTIONS

The criminal sanction is an essential element in the enforcement of environmental controls. Regulatory bodies such as the Environment Agency and local authorities are given these powers in circumstances where polluters have failed to comply with environmental regulation. The fact that they are used to a very limited extent, while environmental pollution continues to be a substantial problem, highlights a major failure in the protection of our environment. The fundamental reason for this state of affairs is the traditional approach by enforcement bodies of seeking a collaborative relationship with polluters. As a result, for polluters, the commission of environmental criminal offences does not have the same stigma as do offences in other areas. Whereas most companies would not dream of practising theft or fraudulent falsification of accounts, the breach of a pollution consent is frequently considered an acceptable part of carrying on an industrial process. It is seen as more akin to parking on a yellow line and risking a parking ticket. Given that a conviction can in many cases attract an unlimited fine and/or up to two years' imprisonment, it is clear that this was not the intention of Parliament when enacting the legislation.

A further aspect of this complacency is that although powers exist to prosecute not only companies but also individuals involved, such as directors of companies, these are very rarely used and the criminal record attaches to the company rather than the director responsible. Environmental campaigners see increasing company directors' personal responsibility for pollution as an essential element in campaigning to protect the environment.

An important first step is to put pressure on enforcement agencies to exercise their criminal sanctions. Unfortunately, owing to the low priority given to criminal prosecutions, such bodies are generally underresourced in this area and have a policy to prosecute only as a last resort. If this is compared with prosecutions in other areas, where commission of an offence is very likely to result in prosecution, the unacceptable nature of this policy becomes very clear. In the short term, therefore, it is very difficult to achieve a substantial change in the approach taken by statutory agencies.

The other approach available to campaigners, which has been used to a very limited extent so far, is for environmental groups to bring private prosecutions themselves. These prosecutions can be brought not only against the polluting companies but also against the directors and managers responsible.

The effects of bringing more private prosecutions would be:

• to highlight the fact that unlawful environmental pollution is taking place;

• to reduce the complacency of polluters about the consequences of breaking the law;

• to highlight to all directors and managers responsible for environmental pollution the individual risk of prosecution faced by themselves as well as their companies;

• to embarrass regulators into taking a more proactive approach to prosecutions.

Which offences can be the subject of private prosecution?

Not all criminal offences can be prosecuted privately. A private prosecution can be brought by any person unless the statutory provision creating the offence provides otherwise, and this is the case for certain environmental offences. For example, the planning legislation specifies that only the local planning authority can prosecute for breaches of enforcement and stop notices. Also, much health and safety legislation provides that only the Health and Safety Executive can prosecute. However, there are important environmental offences where there is no restriction on private prosecution, including:

• offences under section 85 of the Water Resources Act 1991 concerning discharges into controlled waters without or in breach of a consent;

- offences under section 23 of the Environmental Protection Act 1990 relating to failures to comply with requirements of the integrated pollution control or local authority air pollution control regime under Part I of the EPA 1990;

- offences under section 33 of the EPA 1990 involving unauthorised or harmful deposit, treatment or disposal of controlled waste and section 34 of the EPA 1990 concerning breaches of the duty of care in respect of controlled waste.

Liability of directors, managers, etc.

Where an offence is committed by a company under the above and many other statutory provisions, a director, manager, secretary or other similar officer will be guilty of the same offence if it can be proved that the offence was committed with his/her consent or connivance or attributable to any neglect on his/her part.

Such prosecutions can relate to small companies, such as a small skip-hire company where the driver of the skip lorry responsible for the illegal fly tipping is also a director of the company. This is the kind of limited circumstances in which the prosecution provisions are generally used by the statutory authority. They can also be used, however, against directors of large public companies. If the company is responsible for continued illegal discharge from a industrial plant into controlled waters, a director of the company can be criminally liable for the offence. The difficulty will be to prove his/her personal involvement. In such circumstances it is suggested that a letter should be written to the director with responsibility for the environment or other appropriate area, advising him/her of the breach of consent and warning that a prosecution may be brought if it does not stop. It will then be much easier to show, in the event of continuing unlawful discharge, that the director should also be convicted along with the company.

What is involved in a private prosecution?

There are a number of important differences between criminal sanctions and others mentioned in this book. First, whereas in civil actions the person or group bringing the action needs to prove the case 'on the balance of probabilities', in criminal actions the prosecutor must prove the elements of the offence 'beyond reasonable doubt'.

The second difference is that it is usually necessary to prove a criminal intention, or *'mens rea'*, on the part of the defendant – that is, that the defendant not only did the criminal act but also intended to do it. However, this problem does not arise in many environmental prosecutions, as the offences are strict liability offences, meaning that no *mens rea* is required for the defendant to be convicted of the offence.

Third, because of the serious implications for the defendant of being convicted of a criminal offence, the requirements on the prosecutor are more stringent than on a plaintiff in a civil action. It is therefore essential to ensure that lawyers with experience of such prosecutions are involved in preparing the case from a very early stage. Failure to do so may result in a prosecution being thrown out by the court.

Fourth, whereas in all civil actions it is necessary to establish standing to be able to bring an action, no such requirements apply to private prosecutions. Anybody can bring a prosecution in cases where private prosecutions are possible.

Procedure

Criminal cases are heard in the magistrates' court or the Crown Court. Offences that are triable only in the magistrates' court are known as 'summary offences'. Those triable only in the Crown Court are known as 'indictable offences' (none of the environmental offences considered here fall in to this category). The third group into which category most environment offences fall are known as 'offences triable either way', which can be heard either in the magistrates' court or in the Crown Court. Determining which court will try the offence depends on a number of factors, including the seriousness of the offence and the sentencing power of the courts. In the magistrates' court the case will be heard by magistrates, either lay magistrates or a lawyer appointed as a stipendiary magistrate. In the

Crown Court, innocence or guilt will be determined by a jury and, if a conviction is achieved, the judge will then determine the appropriate sentence for the defendant.

Costs

As with all legal actions brought by environmental campaigners, funding the action is an important question. The rules relating to costs in criminal proceedings are very different, however, from those in the other types of action covered in this book. Significantly, there is no power to grant legal aid for a private prosecution (although it may be granted to defend an appeal). (See Chapter 3 on funding.)

Prosecution costs

The court can order that prosecution costs be paid out of central funds, whether or not there is a conviction. The amount paid is at the discretion of the court and it is difficult to predict how much it will be. Private prosecutors' costs are the subject of a practice direction reference [1991] 1 WLR 498, where the Court of Appeal said that a private prosecutor should have costs paid out of public funds except where there is good reason for not doing so – for example, where the case was brought without good cause.

Environmental campaigners who are not receiving legal aid are therefore in a much better position if they take criminal proceedings rather than civil action. The only substantial disadvantage is that the costs will not be awarded until the end of the case, and there is also the very slight risk that costs may be awarded against them. Prosecutors' costs can also be awarded against a convicted defendant to the extent to which they are just and reasonable (section 18(1) of the Prosecution of Offenders Act 1985).

Expenses for witnesses attending court are payable out of central funds but, apart from those for 'experts', these are not particularly generous rates.

Defence costs

A prosecutor can have the defendant's costs awarded against him/her if those costs were incurred by an unnecessary or improper act or omission by the prosecutor. It is therefore essential to take care at every stage to have appropriate legal advice and to conduct the proceedings properly. Provided this is done, costs do not follow the event. This means that an unsuccessful prosecutor will not be liable for the defendant's costs unless any or all of the action was unnecessary or improper.

Penalties

Following conviction, the court has powers to fine or imprison a defendant up to the maximum for each offence as laid down by statutes. In most cases considered here, the maximum for each offence is:

- in the magistrates' court, up to a £20,000 fine and in some cases a three- to six-month term of imprisonment;

- in the Crown Court, an unlimited fine and two years' imprisonment.

Clearly, a company cannot be imprisoned, but directors, etc., convicted of these offences are liable to prison sentences as well as to fines. The courts also have powers to order convicted defendants to pay compensation for personal injury, loss or damage arising from the offence. In the magistrates' court, compensation orders are limited to £5,000 per offence.

Appeals

A right of appeal by the defendant against conviction and/or sentence in the magistrates' court lies to the Crown Court and from the Crown Court to the Court of Appeal.

CONCLUSION

The material dealt with in this chapter is extremely complex, both legally and technically. Unfortunately, environmental campaigners and their advisers have much more limited expertise in these fields than in planning or personal injury and nuisance actions. The areas of dispute tend to be those at issue between regulators and polluters. As a result, most specialist technical experts and lawyers are identified with polluters. To a large extent, polluters are therefore given excessive leeway and flexibility in the application of the law.

In order to achieve greater environmental protection, it is therefore essential for campaigners to establish a greater presence in this area. Over time, with assistance from sympathetic experts and lawyers, this balance can be shifted in favour of the environment.

Part III

ACTIONS TO CLAIM FROM EXISTING DEVELOPMENTS

12 Personal Injury

The most significant feature of pursuing a compensation claim, where it is alleged that environmental pollution caused personal injury, is that it empowers the injured individual to take on the corporate giants who are thought to have caused the illness, in the courts.

Often the illnesses caused by pollution are extremely serious, ranging from leukaemias through to unusual illnesses such as soft tissue sarcomas, where death hangs over the individual and the family for weeks, months and often years. The need for that family to find out what happened and to obtain justice against those possibly to blame is compelling and will often mean the family is unable to live normally until the case has been resolved.

Personal injury claims are not just about the individual's claim. They are also, undoubtedly, one of the most significant ways of bringing to book the polluting company or environmental offender. This type of claim is part of the weaponry for those seeking to protect the environment and, despite its limitations, can be a powerful way of attacking industry.

Such an action can affect the political and regulatory climate and may achieve change. It will also often highlight the concerns of communities over such issues as chemical poisoning, and provide a significant focus for the media interested in reporting such concerns.

This chapter is divided into four main sections. The first section considers the process of proving a case, explaining what the plaintiff needs to prove to win an injury claim. The second examines the court process, identifying the steps that are taken to take the case from commencing the proceedings to trial. The third looks at the specific issues that need to be dealt with during the court process. The final section makes reference to a number of claims that have already gone through or are going through the court process.

PROVING A CASE

For a personal injury claim to succeed the plaintiff must prove that:

- the defendant company owes a duty of care to the plaintiff (see pages 291–4 below);

- the duty has been breached;

- the breach of that duty has caused the injury and damage for which the plaintiff is claiming.

If a plaintiff can establish these three points in a court of law, then negligence or breach of statutory duty may be established and a finding of liability will be made by the judge, who will go on to assess the value or 'quantum' of the claim – that is, the compensatory award of damages.

In the British courts the burden falls squarely on the shoulders of the plaintiff to prove the case that is being brought. It is not for the defendant to disprove anything, and indeed in a trial it is open to the defendant to ask the court to give judgment against the plaintiff before even presenting his/her evidence, if the court is satisfied there is no case for the defendant to answer.

There are, therefore, two primary hurdles for the plaintiff in these actions: first, proving that the defendant had been negligent or in breach of a statutory duty, and second, showing that the actions of the defendant caused the injury.

Duties and breaches

As has been described in previous chapters, the rule of law in Britain has two arms: laws determined by Parliament, that is, statutory law; and the common law, by which is meant laws determined by the courts over the years through judgments made in the many cases that appear before them every day.

The role of the courts is to follow the laws established by Parliament, but there are many areas of life where the Houses of Parliament have not prepared laws or have drafted laws open to differing interpretation, and it is in these areas where judge-made law or the common law operates.

In a personal injury case the first issue that needs to be established is whether, in the circumstances that arise, there is a duty in statutory law or the common law that applies to the defendants and that they might have breached.

Statutory duty

There are a number of duties that have been imposed by Parliament regarding the exposure of populations to pollution. The duties are not at all general but are very specific to each industry. Where a person has been allegedly injured by some sort of occurrence of pollution, it is therefore important to consider the specific legislation controlling the pollution emanating from a particular plant or process. It is impossible to be exhaustive here but a few key examples are given.

Radiation

Section 7 of the Nuclear Installations Act 1965 says very specifically that where anyone can show that an injury, or indeed any damage to property, has resulted from exposure to radiation emanating from a particular plant then that person will be compensated without the need to show that a duty has actually been breached. This concept is known as 'strict liability'. It is very significant because it effectively eradicates the hurdle of showing the defendants to the action have breached a particular duty.

The concept of strict liability is being considered by the European Commission for extension to other areas of environmental pollution, but because of the enormous lobbying by industry, such extension has not as yet emerged. It is likely, however, to become a more widely available principle in years to come.

Sewage

The discharge of sewage and other pollutants into our rivers and seas is strictly governed by a series of 'consents', issued formerly by the National Rivers Authority (NRA), and since April 1996 by the Environment Agency. The consents very specifically state what can and cannot be spewed into the rivers and seas and what treatment of sewage needs to take place before it is allowed out into the waterways of Britain.

Any case involving injury to people using the rivers and seas would require a close look at these consents to see if they had been breached (see Chapter 15 for more on cases involving sewage contamination).

Industrial plants

The aerial release of chemicals is strictly governed, primarily by the former HMIP (Her Majesty's Inspectorate of Pollution) now the Environment Agency, and is largely monitored by local authorities. Again, therefore, where injuries are alleged to have occurred around these facilities, a close inspection of the authorisations given for these releases and the monitoring of them will be very important.

Pesticides

Pesticides are subject to special government controls under an array of legislation including the Food and Environment Protection Act 1985 (FEPA) and the Control of Pesticides Regulations 1986 (COPR). Pesticides are deemed safe by the authorisation and licence of six government department ministers who are advised by a body known as the Advisory Committee on Pesticides (ACP). Many 'older' pesticides have not been fully evaluated as to health risks but continue to be sold.

For those who live within the vicinity of farms or other areas where pesticides are sprayed onto fields, woodlands, etc., there is always the prospect of inhaling or otherwise coming into contact with the chemicals with serious injuries resulting (a phenomenon often called spray drift). In these circumstances, breaches of authorisations should be considered as should the Health and Safety Executive (HSE) publications, including the agricultural codes of good practice relating to the application of pesticides. These may also be applicable where a company manufacturing chemicals causes a chemical spill in the locality.

Worker cases

There may be a relationship between exposure of workers and exposure of people living within the immediate vicinity of the same plant. At Sellafield, for example, there are wire fences to prevent people breaking into the plant. The fact that the radiation workers at the plant can be exposed to high doses of radiation means that those living in the immediate vicinity of the nuclear facility are also likely to experience the impact of the plant's radioactivity. A wire fence is going to do little to protect the public in these circumstances.

There is a vast array of specific legislation relating to health and safety at work, including the Health and Safety at Work Act 1974, the Factories Act 1961 and specific industry-related regulations, such as those for asbestos, shipbuilding, power presses and chrome plating. Also there is the protection of workers afforded by European Union directives, including minimum health and safety requirements for the use by workers of personal protective equipment in the workplace. The standard often quoted is BATNEEC, 'best available technology not entailing excessive cost', but some cynics remark that CATNIPP is more often invoked – that is, 'cheapest available technology not involving prolonged publicity'.

The manufacture and use of poisonous chemicals are strictly controlled by the Control of Substances Hazardous to Health Regulations 1988 (COSHH). An employer must assess the risks of hazardous chemicals and take steps to substitute them.

Correct provision of appropriate personal and respiratory protective equipment (i.e. protective clothing and masks worn to protect workers from contact with chemicals) will be of paramount importance for employers to avoid actions in negligence.

Consumer product liability

Consumer product liability may well become important in establishing liability of manufacturers, particularly in relation to chemical/pesticide products (including sheep-dips). There is a strict liability (liability without the consumer having to prove fault) under the Consumer Protection Act 1987 for products that cause injury or harm. The scope of this Act may well be significantly impacted upon by new and existing European directives.

Common-law duty

The courts have for many years imposed a general duty of care on each of us in society to have due regard for our neighbours. This 'neighbour' principle is the cornerstone for most of the claims within this field, and the standard expected was set out by Lord Atkin in 1932 in the classic case of *Donoghue* v. *Stevenson*:

> You must take reasonable care to avoid acts or omissions which you can reasonably foresee would be likely to injure your neighbour. Who then, in law, is my neighbour? The answer seems to be – persons who are so closely and directly affected by my act that I ought reasonably to have them in contemplation as being so affected when I am directing my mind to the acts or omissions which are called into question.

Subsequent courts have defined these words as meaning that there is a duty where it can be reasonably foreseen that injury might arise from the person or company's action to people reasonably close to what is happening and where they are directly affected.

This issue of foreseeability has become central to this duty. In the Cambridge Water case, described in more detail in Chapter 13 (page 336) and Chapter 14 (page 352), where damage was caused as a result of chemicals seeping through into the ground water from the defendant's premises, the House of Lords held that because this was not foreseeable the defendants were not liable for the damage caused. Although this was primarily a nuisance action, it gives a perspective on how narrowly the courts are likely to deal with this issue.

In personal injury claims, therefore, it is often difficult to prove that a defendant was aware or indeed should have been aware of the potential of its chemical(s) to cause a specific illness. A manufacturer or user of chemicals will argue that the medical and scientific evidence was not widely available, that it was found in often obscure medical and scientific journals. The company will ask how could it be expected to have knowledge without considerable research, and without taking advice from experts in subjects relating to rare diseases. This is called the 'state of the art' defence – that is, at the time they could not have been expected to know.

In clarifying this foreseeability test one judge remarked, in the case of *Glasgow Corporation* v. *Muir* (1943):

> unless the needle that measures probability of the particular result flowing from the conduct of a human agent is near the top of the scale, it may be hard to conclude that it has risen sufficiently from the bottom to create the duty reasonably to foresee it.

Alongside the foreseeability standard, where a product itself can cause harm – say, cigarettes through passive smoking – another correlating duty imposed on industry by the courts relevant to environmental personal injury cases is that set out in *Wright* v. *Dunlop Rubber Co. Ltd* (1971). Here the court stated that where a product is found to have harmful effects the producers must take reasonable steps to eliminate – or, at worst, materially reduce – those effects. Further, the courts suggested that the duty to consider the impact of their products was a continuing one and not solely a duty imposed when the product was initially manufactured. The court said:

> It is obvious that the answer to the question 'What are reasonable steps?' must depend upon the particular facts. It is

obvious also that the duty is not necessarily confined to the period before the product is first produced or put on the market. Thus if when a product is first marketed there is no reason to suppose that it is carcinogenic, the manufacturer has failed in his duty if he has failed to do whatever may have been reasonable in the circumstances in keeping up to date with the knowledge of such developments and acting with whatever promptness fairly reflects the nature of the information and the seriousness of the possible consequences. If the manufacturer discovers that the product is unsafe, or has reason to believe that it may be unsafe, his duty may be to cease forthwith to manufacture or supply the product in its unsafe form.

The judge went on to say that in some circumstances warnings may be sufficient but in grave circumstances withdrawal of the product may be necessary.

The courts have accepted that some manufacturing processes are so hazardous as to be 'dangerous *per se*' – that is, they are in a category of exceptional risk, hazard or danger – and in those instances the courts are prepared to impose a qualified strict liability upon the manufacturers. This view has been whittled down over the years to Lord McMillan's comments in the case of *Read* v. *J. Lyons & Co. Ltd* (1947):

Accordingly, I am unable to accept the proposition that in law the manufacture of high-explosive shells is a dangerous operation which imposes on the manufacturer an absolute liability for any personal injuries which may be sustained in consequence of his operations. Strict liability, if you will, is imposed upon him in the sense that he must exercise a high degree of care but that is all. The sound view, in my opinion, is that the law in all cases exacts a degree of care commensurate with the risk created.

The last line of the judge's comment possibly best sums up the court's overall approach.

There are occasions when the common law and statutory law work in parallel and there are other times when they can conflict.

For example, the fact that statutory law allows people to drive at 70 m.p.h. down the motorway does not mean that if they are driving carelessly and cause an accident at that or lower speed they cannot be successfully sued for breaching the standard of care expected of them.

However, in *Budden* v. *BP Oil Ltd* (unreported 1980) case, the laws were in conflict. Here, in the late 1970s, the parents of children who lived close to the M4 motorway were claiming that, as a result of the high levels of lead in petrol, the exhaust fumes were causing illness in their offspring. Their claim was that the petrol manufacturers were breaching their common-law duty of care to those living in the vicinity of motorways, including their children.

Although the lead levels in petrol may have been in breach of the common law, the fact was that there was a statutory maximum level of lead in petrol permitted by Parliament, and it was accepted that the manufacturers had complied with these requirements. The court essentially took the view that the statutory framework took precedence and that providing the manufacturer complied with regulations then such legal action would fail.

Causation

This is, without doubt, the largest hurdle faced by the plaintiffs in most personal injury actions.

It is very important for the potential plaintiff to be aware that a seemingly apparent link is not sufficient to satisfy the standard of proof required by a civil court of law. The plaintiff must show that but for the action of the defendant the injury and damage would not have been suffered. This is called the 'but for' test.

The leading case describing the tests applied by the courts in terms of showing proof of causation by a defendant is the standard set by the House of Lords in *Wilsher* v. *Essex Area Health Authority* (1988). In that case, where a child suffered from severe deformities at birth, there were various possible causes of the illness, one of which was that the hospital where the child had been born was negligent. The presiding Law Lord, Lord Bridge, in considering the evidence, stated that the plaintiff had to show that the act complained of had either caused or materially contributed to the onset of the injury or illness. If there are just as likely – or indeed more likely – alternative, competing explanations for the injury other than that complained of, as there were in that case, then the plaintiff will fail to prove the case. In that case the infant plaintiff failed to meet this standard.

Cases of environmental pollution are notoriously difficult to prove, whether due to chemical poisoning, radiation, electromagnetic fields, or whatever. The primary reason for this is the difficulty in showing that the illness was caused by the particular pollutant, an essential part of an injury claim.

In many other personal injury claims causation is not particularly difficult to prove. In a car crash, for example, where the plaintiff was previously well, and immediately after suffers from, say, a broken leg, bruising and serious trauma, each is readily tied in to the accident. In environmental claims this immediate relationship is not at all apparent.

Clearly there are instances where the relationship is clear. For example, when, in 1988, a lorry accidentally tipped its load of aluminium sulphide compound into the water supply of the people of Camelford, there was little difficulty in showing that the immediately arising sicknesses were caused by the polluted water. What is far more difficult is to show that pollution over a longer period of exposure (known as chronic exposure) has caused a particular illness where that illness arises in the wider population.

For example, in the early 1980s it was discovered that there was a tenfold excess of childhood leukaemias in the village of Seascale, only two miles from the Sellafield nuclear reprocessing plant. It is generally accepted that radiation is one of the only known causes of leukaemia, and that Sellafield has discharged into the air and sea more radioactive waste than any other nuclear facility in the western world.

Despite it seeming on the face of it that the plant must be to blame, the fact that childhood leukaemia is the most common childhood cancer meant that in the ensuing claims the plaintiffs' lawyers had to try to disentangle the prospect that there might be other competing causes of the cancers. In the end, the judge in the case came to the view that, although radiation was a possible cause, there were other possibilities and that the standard of proof had not been reached.

To understand the processes involved in proving this most difficult of issues of causation it will help to consider how the courts approach proof of causation in more detail.

In the unreported case of *Gaskill* v. *Rentokil* in 1994, the trial judge, Mr Justice Otton (now Lord Justice Otton), suggested the following approach. The case concerned an allegation that the child, William Gaskill, had contracted a rare blood disease, called aplastic anaemia, as a result of the spraying, on two separate occasions, of his home with lindane. Lindane (gamma hexachlorocyclohexane – Gamma HCH) is an organochlorine insecticide and pesticide used in the treatment of wood for woodworm.

The defendants were Rentokil, the well-known British wood-treatment company using pesticides. Their insurers had previously settled out of court the claim of another child whose bedroom had been sprayed with the same substance. That child had suffered from the same illness and had subsequently died.

The judge asked himself the following three questions:

1 Can lindane (the chemical) cause aplastic anaemia in humans?
This is what lawyers call 'generic causation' – that is, is the chemical deemed generally capable of causing this particular injury or illness? This is primarily a scientific issue.

The first problem for the judge and the parties is that the medical and scientific community will generally be looking at scientific standards of proof in making their assessment as to whether A *causes* B. It is clear that this standard is a lot higher than the balance of probabilities. Reports and articles in the medico-scientific literature will merely indicate a possible 'association', often illustrated by case histories. This does not establish cause to scientific standards. An 'association' is exactly what it means in plain English. It is not scientific and medical proof but merely evidence that an association between one factor and another has been noted, often in case histories or epidemiology. Epidemiology is the statistical study of the cause of illnesses and the attempt to establish a relationship between illnesses and specific possible causes. A classic epidemiology study is that of Professor Sir Richard Doll and his colleagues in which they concluded that smoking can cause cancer.

The medical and scientific community is often divided as to whether A causes B. As an example, in one case two medical experts of considerable stature, an epidemiologist and an oncologist (a doctor specialising in cancers), disagreed whether pentachlorophenol (a wood-preservative) was capable of causing soft tissue sarcoma, the epidemiologist saying it was around 50 per cent likely to be true, and the oncologist suggesting the likelihood of it being true was nearer 85 per cent.

2 Was the plaintiff exposed to lindane (the chemical) and if so to what extent?

This deals with the dose of the chemical at the time of exposure. It is important to remember the words of Paracelsus (1567) that 'all substances are poisons, there is none which is not a poison. The right dose differentiates a poison from a remedy.' However, the issue of idiosyncrasy and chemical sensitivity must be considered.

All environmental personal injury claims require expert evidence as to the effect of a particular chemical or substance on health. Modern toxicology is defined as the study of the adverse effects of xenobiotics (toxic agents). Ever since the work of the chemist Paracelsus in the fifteenth century, dose has been the mainstay of understanding and assessment in the field of toxicology.

The toxicologist will consider any adverse effect by considering, among other things, the duration and frequency of dose. Models have thus been developed, including Lethal Dose (LD) values. Lack of space here prevents a full discussion. An excellent textbook is Casarett and Doull, *Toxicology – The Basic Science of Poisons* (1992). It is worth pointing out that a medicinal application for lindane was its use in low doses to treat the parasitic condition scabies, and as a head lice shampoo. Thus a small dose of a chemical may be therapeutic as a remedy for a disease, whereas an excessive dose of the same chemical may cause injury.

3 Did lindane (the chemical) cause the plaintiff's aplastic anaemia (the injury)?

This is called 'specific' causation, which differs from 'generic' causation in that the individual circumstances are considered on a case-by-case basis. There can be other reasons why an individual contracts aplastic anaemia (or other injury or illness): for example, genetic cause or predisposition, or virus. Thus the competing likely causes for the injury may outweigh the allegation brought against the defendant that its use of chemicals was the cause. In the Gaskill case the action failed, the judge finding that, as the onset of the disease was more than 18 months after the child's exposure, lindane was a possible but not proven candidate. Also, in specific causation the individual's past medical history will have to be considered.

The standard of proof

The hurdles that the plaintiff has to overcome, in terms of duty of care, breach and showing the breach caused the injury, have been described. The next significant question for the potential plaintiff is the level of proof for each of these hurdles that the judge will require.

In criminal cases the prosecution has to show that the defendant is guilty 'beyond reasonable doubt', which is a standard that is very high. However, the legal standard of proof in civil cases, such as environmental personal injury, is much lower. All that is required is that the plaintiff should prove the case 'on the balance of probabilities', which can also be read as being 'more likely than not'.

The borderline for these types of case is, therefore, only 51 per cent, which is a lot lower than that required by the prosecution in order to secure a conviction in a criminal trial. A plaintiff's claim will succeed, therefore, if the case is proved beyond mere suspicion and to a level of more than 50 per cent likely.

Having stated the theory, the point needs to be made that experience of bringing this type of environmental litigation carries a perceived feeling of a reluctance by the judiciary to find for the plaintiffs in these types of claim. This may be partly because this standard of proof is difficult to overcome, but is also perhaps because the judiciary is wary of opening the floodgates for further litigation and may also find that decisions at this level are extremely politically sensitive. It takes a brave judge to come to the view that a major British company is causing pollution to the environment that is causing widespread serious illness in the local population.

At times it would appear that the courts are looking at a stricter level of proof than the balance of probabilities.

THE COURT PROCESS

Having considered the main hurdles to be overcome in a personal injury action, and the standard of proof at which the lawyers will be aiming, it is important to turn to the formal stages of the legal process so that the potential plaintiff can be clear as to the likely route the case will take on its way to a court determining the issues.

The pleadings

In any legal action there is a set of formal court documents that set out the claim and the response, which are known as the 'pleadings'. The legal action for damages is commenced by the issue of a writ and statement of claim, which set out the facts and allegations of the plaintiff's claim. The defendant acknowledges those proceedings and then serves a defence, which is supposed to set out exactly where the defendant stands in relation to each of the allegations made in the statement of claim. The reality, however, is that the defence is usually a series of denials of – or, at best, refusals to admit – the claim. It rarely takes the case much further.

Very often the defendants will serve a document called a request for further and better particulars of the statement of claim, usually at the same time as the defence is served. This document seeks to make the plaintiff give further detail about the claim that is being made.

Occasionally, the plaintiff's lawyers will serve a reply to the defence and/or a request for further and better particulars of the defence. Usually this is not necessary.

Directions

The next step in the action is known as the directions stage, which is about the establishment of a timetable and procedure for the disposal of the case. This is determined by a court order. In less complex cases this is automatic and the terms of it are standard: that is, the parties to the action are bound by a predetermined set of directions. In more complex cases the directions are far more flexible and usually one side or the other issues a request to the court, a summons for directions, which sets out the suggested order and which has a time and date allocated for the two sides to turn up at court for the issue to be determined by the court, to the extent that agreement has not been reached.

The directions order deals with the dates for the exchange of the experts reports, both medical and non-medical (for example by a consulting engineer). Also witness statements have to be exchanged from all the witnesses as to fact at any trial. This order will deal with the timescale for this to take place and will also usually deal with the timetable for the setting of the action down for trial – that is, asking the court for trial date – and, perhaps most important, will deal with discovery.

Discovery

A key issue in environmental pollution cases is that of discovery. What is meant by this is that each side is supposed to produce to the other side all documents in its custody, power or possession that relate to the issues in the action. These documents can include confidential or commercially sensitive information from chemical companies, if this is relevant to the issue at trial. The prospect of a successful action against a polluter may well be determined by the manner and amount of discovery of documents. In the US, incriminating documents belonging to the defendant are sometimes called 'hot documents' or the 'smoking gun' – that is, those that give clear support for the negligent act being alleged.

The sorts of documents that may be relevant could include internal memos abut the development and design of products, correspondence with regulators such as the Health and Safety Executive, records of complaints or defect reports, monitoring records, etc.

Discovery is commenced by both parties listing all the relevant documents that are (or have been) in their possession, custody or control (or the control of others) in a 'list of documents'. Sometimes, one of the parties to the action might not be entirely frank about these documents and a listed document may indicate the existence of further documents not on the list. The role of the plaintiff's lawyer is to make this assessment and to request the revealed documents. Thus the paper chase is under way.

In one case, an important document – a 'code of practice' (that is, the industry's good work practice) – was missing from the defendant's list and also from the trade association library, the British Library and the government regulator's file. Upon enquiry a government staff employee said, 'Oh, that file has been removed.' The document was discovered purely by chance when an expert found an old copy in a previously unrelated litigated case against the same company. It appeared that the code of practice had been deliberately removed from all the usual known sources.

One excuse can be that the company papers have been culled or placed onto microfiche with resultant loss of documents. While companies obviously do have to clear out old files from time to time, it is interesting to note how often it is the 'smoking guns' that appear to have been selected for clearing out.

The plaintiff's lawyers have to take great care, therefore, to consider what is available from a defendant's list of documents, not just for what it contains but often for what it *should* contain.

If it is suspected that either party has been playing fast and loose on discovery, an application for specific discovery can be made. This is a request for specific documents (such as a specific memo thought to exist) or a specific class of documents (for example, all reports of production test results) thought to have been omitted. The party opposing the application must then produce an affidavit as to the veracity and accuracy of the original list of documents. The courts can impose penalties for failure to observe proper and truthful discovery. In extreme cases, the court can strike out the case of the offending party.

Also of importance are third party or non-party discovery applications. This involves obtaining documentation from those not actually sued. This can include the regulatory authorities, such as the Health and Safety Executive, National Radiological Protection Board or Veterinary Medicines Directorate, or any other organisation that has relevant documentary evidence. However, the party requesting the documents usually has to pay the legal costs of the non-party.

Another important weapon for the injured party's lawyer is that of pre-action discovery. This can be crucial in chemical poisoning and other such cases where it is vital to ascertain the toxin involved, and to obtain as much information as possible about the chemical or product used at the time of exposure. Safety data sheets often assist.

Pre-action discovery proceedings can be commenced pursuant to section 33(2) of the Supreme Court Act 1981 and Order 24 Rule 7(a) of the Rules of the Supreme Court 1965. The sections do, however, only allow for such a process to take place against the party that is about to be sued. It is not possible to obtain pre-action discovery against a non-party to the action.

This process is of great importance, as the documents may assist an expert in advising or showing that there is no case, thus preventing the waste in time and costs of fighting a hopeless case. Although the party requesting the documents usually has to pay the legal costs of the other party, this is not always the case. In the case of *R*. v. *Rentokil* (1988), in a pre-action discovery application following a refusal by Rentokil Limited and its insurers to provide information or documents about the active ingredients, pesticides and solvents in their products at this early stage, the court not only awarded discovery but also ordered Rentokil Limited to pay the substantial legal costs of the application as well.

Other procedures and documents

There are various other significant legal procedures that are open for either side to use. These include 'interrogatories' and 'notices to admit'. Interrogatories are written questions from either party that require specific answers. Notices to admit facts are usually an attempt by one of the parties to narrow down the scope of the issues to be determined by the trial judge by asking the other side to admit certain facts. If they refuse and if the judge takes the view that the facts should have been admitted, then even if that side wins the case the judge can order that they pay the costs of that issue being determined.

The trial

The court will usually have read all of the formal documents in the case before the trial commences and these days, with the pressure being for trials to take less and less time, there is an increasing tendency for most issues to be dealt with by paperwork.

The date for the trial in complex environmental claims is usually fixed, which provides an element of certainty about the starting date. This compares favourably with many shorter cases, where the case may be listed for a particular date but there is no certainty of it being heard then. The reason for this treatment is that environmental cases usually involve complex issues requiring senior scientists and experts to give evidence. Their time is seen as being extremely expensive and valuable and not to be wasted in their hanging around to give evidence at court.

The trial takes place before a single judge. The format is usually that the plaintiff presents his/her case first, followed by the defendant. The case commences with the plaintiff's barrister, also known as counsel, presenting the trial judge with a broad picture of the facts, the law and generally how the plaintiff sees the case. This opening statement can be followed by a response from the defendant's counsel.

Once this stage is over, the plaintiff is usually called first to give evidence in the witness box, followed by witnesses as to fact and then the experts. The defendants then set out their case in a similar way by calling witnesses and experts. Finally, the barristers for both sides give closing statements, with those for the defence going first, setting out how they consider the evidence has gone and their view as to what the judge should find.

It is then up to the judge to give a judgment. In more complex cases this will usually entail his/her retiring, sometimes for weeks and maybe even months, before giving his/her decision. The judgment is handed down (that is, read out) in open court by the judge, although these days it will often have been typed and the typed pages distributed to the two sides' lawyers a day or two before, which may mean it is not read out on the day but simply handed to the media available at court.

The judge will then deal with the issue of the legal costs of the action (see Chapter 3) and any requests for authority to appeal.

Appeals

Appeals are heard by the Court of Appeal and rarely by appeal again to the House of Lords, the final court for appeals. These courts will only hear an appeal against the trial judge's decision where it can be shown that there is an error of law, or that the judge has been unreasonable in some way as to the conduct of the trial.

It is fair to say that an extraordinarily small number of personal injury cases end up reaching the appeal stage. This is because this area of law is fairly well established and, apart from specific procedural points, little new law tends to evolve.

SPECIFIC ISSUES

In addition to the various hurdles set out above, in these cases there are other problems that the victim has to take into account.

Limitations

One of the most important issues that victims need to be aware of is that the courts impose time limits within which a claim for personal injuries can be brought. A plaintiff must issue proceedings in the High Court or county court within three years of knowing they have a claim. If the plaintiff delays longer than this, unless there is a compelling reason for the court to exercise its discretion, the claim is likely to be struck out as 'statute barred'. The court will not allow the plaintiff to bring the action after such a delay.

For a child, the three years only start to run from his/her eighteenth birthday, unless the child dies, in which case they run from the time of death. Time does not run for a person who is mentally incapable – and therefore does not have legal capacity –

unless and until that capacity has returned (see Chapter 9, page 194, for more on capacity).

The date of knowledge – that is, when the court attributes knowledge to the potential plaintiff – is when the plaintiff knew the injury in question was significant – that is, sufficiently serious to justify action against a non-contesting defendant – and attributable in whole or in part to the act or omission that is to be criticised. A plaintiff may be deemed to have knowledge – that is, to have 'constructive' knowledge of facts that could and should have been discovered by the plaintiff, with or without expert advice that the plaintiff could be reasonably expected to get. The courts are in flux and this will be interpreted more or less flexibly by different judges. It is certainly not safe to presume that the limitation clock has not started ticking until a medical or other specialist has confirmed the plaintiff's suspicion that the injury was caused by some sort of pollutant.

The issue can be complicated by factors such as whether an individual was too ill or disabled to appreciate injury, whether the individual and even the doctors expected the injury to be temporary, and when was a reasonable time for a plaintiff to start making enquiries. The time will not run from when the plaintiff should have begun to make enquiries because the court will allow a reasonable time for actual knowledge to be obtained following that. The safest rule, however, is for the victim to consult a solicitor as soon as it is suspected that a claim might be feasible. Delay can be fatal to any chances of recovering damages.

Locating experts to support a claim

Locating experts can be one of the most difficult but often most rewarding aspects of personal injury litigation. In the field of environmentally related injuries, it is extremely difficult to locate high-quality medical specialists with either the inclination or the time to devote to compensation claims. It is vital to locate the right expert, especially in relation to proving medical causation. Although judges listen politely to non-medical experts in biological science and other academic disciplines, it is noteworthy that a judge's pen moves with extra speed when medical evidence is given by a medic.

Without medical and scientific evidence to support a personal injury claim there is very little chance that it will succeed, unless the injuries are minimal and events straightforward.

It is the job of the plaintiff's lawyers (as it is for their counterparts on the defendants' side) to obtain the assistance of eminent experts who take a view that is compatible with their side's case. Quite often there are experts to whom the plaintiff has turned during the early part of his/her taking action and whom the plaintiff is keen to have acting on his/her side. However, more often than not these experts are quasi-campaigners, and may well not be the most appropriate people to give evidence at the trial. It is important that the plaintiff understands this point, as it can be the case that the plaintiff and the lawyer start pulling in opposite directions.

Another point the lawyer will be looking for is to ensure that the experts prepare reports that cover only their own area of expertise. It is very easy for an expert to stray into a peripheral area of the case, thinking that his/her general knowledge of the issue is sufficient to give an opinion. That is usually not the case, however, and the plaintiff's lawyer will be trying to ensure that there are appropriate numbers of experts to cover the primary areas at issue so that they can readily hold their own when it comes to cross-examination by the defence counsel.

In locating experts it is important for the lawyer not to look solely to those in the UK. Often, experts overseas have far more experience in certain areas than those in the UK. For example, the Scandinavians have carried out all of the ground-breaking work in solvent-induced injury, including chronic encephalopathy (cerebral injury without lesion), there being little research in the UK. In the area of electromagnetic fields and the hypothesised link with various cancers, very little work has taken place in Britain. Most of it has occurred in either the US or Scandinavia and, therefore, trying to locate purely British experts would undoubtedly seriously weaken the strength of the case in either instance.

The only caveat is that British judges can be both sceptical and critical if there are too many foreign experts, as there is still an unstated view among some judges that British experts are likely to be the best and most reliable.

A useful way of tracking down an expert is to undertake a scientific or medical database search such as 'Medline'. From this, copies of particular papers of relevance can be obtained from the British Library or other sources, which will usually give the authors' hospital or academic addresses, and this is the necessary trail for locating the right experts. Eventually one expert will recommend another and, no matter how obscure the point, there is probably someone out there who has studied it for years.

Damages and settlements

The compensatory awards of damages (or 'quantum') for personal injuries are far lower than most people would consider reasonable. The general approach of the court is to place the injured plaintiff in the same financial position as if the accident had not occurred. The courts accept and recognise that no amount of money awarded by way of damages can ever replace an actual bodily injury or truly compensate for the loss of a relative, so awards of damages can only reflect a judicial view of a financial substitute.

Damages awards are divided into two main categories or 'heads of damage':

• special damages – for financial loss that has occurred up to the date of trial, including loss of income, and for future financial losses, again including loss of income and care costs;

• general damages – for pain and suffering and loss of amenities: that is, for the injuries themselves.

Much publicity is generated within the media when million-pound awards are made, but the reality is that such awards are rare and usually involve claims relating to brain-damaged claimants who need constant care for the remainder of their lives. The award in such a case will often include many years of lost financial income, future costs of professional carers and substantial adaptations to the home.

It is important, therefore, that potential plaintiffs do not rely on their reading about these kinds of case to gain a feel for what their cases are worth. Such a view is only likely to lead to severe disappointment when and if the claim is resolved.

Pain and suffering awards
It is important to have some idea as to the scale of damage awards in this country. The Judicial Studies Board's *Guidelines for the Assessment of General Damages in Personal Injury Cases* (1994) gives guidance to judges as to general damages awards for pain and suffering, examples being as follows:

Total blindness and deafness	£125,000
Very severe brain damage – vegetative state	£105,000 to £125,000
Loss of both arms	£85,000 to £95,000
Below elbow amputation	£37,500 to £42,500
Total deafness	£35,000 to £42,500
Complete loss of sight in one eye	£20,000 to £22,500
Loss of one kidney (with no damage to the other)	£12,500 to £17,500
Amputation of ring and little fingers	£9,000
Loss of spleen	£5,000 to £8,000
Slight hearing loss/occasional tinnitus	£3,000 to £5,000
Simple fracture of forearm	£2,750 to £8,000
Minor head injury	£1,000 to £5,000
Loss of one front tooth	£1,000 to £1,500

These are 'general damages awards'. Precedents for other aspects of damages awards can be found in various legal journals and textbooks such as Kemp and Kemp, *The Quantum of Damages* (1982), which is constantly updated as a loose-leaf system. With this almost tariff-based system of awards, an experienced personal injury lawyer will have a fairly clear idea of the likely value of the claim at the first interview with the client. This view becomes

fine-tuned as the case moves on towards trial, with a schedule of
financial loss having to be served with the statement of claim and
with a final schedule being served much closer to trial.

Claims for damages of a psychological or neurobehavioural nature
Under the heading of general damages for pain and suffering has
increasingly come claims for post-traumatic stress disorder (PTSD).
The understanding of this type of injury and its development as a
category or 'head' of damages has come about primarily as a result of
tragic events such as the Kings Cross Fire and the Zeebrugge ferry
disaster. In each of these cases there were many victims who suffered
from a variety of stress-related symptoms, where their illnesses went
well beyond simple grief. In both cases victims were given substantial awards under the head of PTSD, and, with a greater understanding of the psychological processes involved, the claim has been extended to other areas, including some environmental claims.

For example, PTSD may well be a consequence following chemical poisoning, where such an event will cause psychological trauma
in addition to the physical effects of the chemical itself. Also in the
specific case of organophosphates (OPs), used in sheep-dips, it has
been established in the medical literature that OPs can cause long-
term psychological or neurobehavioural damage.

In 1996, the House of Lords considered for a second time the
case of *Page* v. *Smith*, and upheld an award of very substantial
damages to a teacher who suffered from myalgic encephalomyelitis
(ME), which was in remission but was triggered by a road traffic
accident. He was not otherwise injured in the road accident. This
shows an increasing tendency by the courts to accept the need to
award damages for these more complex ailments.

Financial losses

Within the schedules there must also be set out in some detail the
financial losses, including the loss of earnings claim, both present
and future, which is usually the largest part of a claim.

Other likely sums to be included within the schedules are the
cost of replacement clothing, travel expenses to and from hospital,
and the cost of medicines and private medical treatment.

In an extreme example, an employed 30-year-old with a good, well-paid job, unable ever to return to work, is likely to recover very substantial damages to the tune of hundreds of thousands of pounds. However, claims for people who have a long-term unemployment history and with no real prospect of working in the future can be rather low.

Private medical expenses can be fully recovered, so that the victim can choose to have private medical care rather than through the NHS. However, investing in private medical care is risky unless a case is very clear cut because, if the case is not won, the victim will not only be left with no compensation but also will have paid a hefty bill for the private health care.

Interest

Interest is recoverable on some of the heads of damage at rates set by the court, with general damages for pain and suffering having interest that accrues from the date of the commencement of the proceedings and with special damages having interest that accrues from the date of the injury. For this reason, in terms of interest, it is beneficial to issue proceedings as soon as possible.

Offers to settle the claim and 'payments into court'

At some point in the litigation a defendant may make an offer to settle the claim, and may also choose to pay into court a sum reflecting that offer. The offer can be the total value of a claim, in which case the lawyers will advise the client to settle. It is important to appreciate that the court, when awarding compensation for an injury, can only award compensatory (monetary) damages. The court, if the plaintiff is fully compensated, will not be concerned with resolving who was responsible. The case concludes with the payment of damages and, if this is on an agreed settlement without formal admission of fault by the defendant, the issue of responsibility or liability will never be resolved.

However, an offer of full compensation is rare in these types of cases and an important card that can be played by the defendant is 'payment into court'. Many litigants who have struggled through the legal system claiming compensation know only too well its deadly effect.

If, for example, a payment into court of £10,000 is made by the defendants, in full and final settlement of the claim, and it is not accepted by the plaintiff either within 21 days or later than that with agreement of the defendant and leave of the court, then all legal costs (those of both sides) incurred from the date of payment in will have to be met by the plaintiff, if the judge subsequently awards a sum less than or the same as the amount paid in. In these circumstances, many lawyers, with all the problems of proof in this type of case, may advise acceptance, even if the figure paid in is much less than the full value of the claim. The judge-at-trial does not know the amount of the payment into court.

If such a settlement opportunity arises and is turned down by the plaintiff, the plaintiff's lawyers will have to advise the Legal Aid Board, trade union, etc. (whoever is funding the legal costs of the action) of the prospects of the litigation succeeding, and in most cases where the lawyer is advising acceptance the funder will insist on the offer being taken, with the threat that if this is not done the funder will withdraw financial support for the case. Undoubtedly plaintiffs in this situation will feel let down, not only at the size of the settlement but often also because they have not had their day in court. Money may not be the main motive of plaintiffs going to law, and it may be that they feel that their case exposes a fault in the system. Plaintiffs undoubtedly feel pressured by offers of settlement through payment into court. However, as the law presently stands, it is almost inevitable that the case will end in this unfortunate way.

EXAMPLES OF INJURY CLAIMS

There have been some successes for victims of environmental pollution but there have also been a number of losses. Without question this is a difficult area for plaintiffs. Set out below are some brief histories of cases that have reached the courts in this field.

Sellafield childhood leukaemia victims

Some ten childhood leukaemia victims, whose fathers primarily worked at the plant, sued British Nuclear Fuels Limited, the operators of the Sellafield nuclear reprocessing plant. The basis of the claim was that radiation from the plant had caused the cancers, either directly through the radioactive discharges or indirectly through occupational damage to their fathers' sperm by the radiation, which subsequently led to their contracting the cancers.

This was a massive case with around 35 experts from all over the world on each side. The case lasted nine months, and generated some half a million pages of documents. Mr Justice French, giving his judgment in October 1993, gave the view that although radiation was a possible cause of the cancers he was not satisfied on the balance of probabilities that it was the cause.

Molinari v. The Ministry of Defence (unreported 1995)

Rudi Molinari worked as a fitter in the Ministry of Defence naval dockyard at Chatham. He was subjected to significant levels of exposure to radiation while working on the refitting of nuclear-powered submarines, following which he was diagnosed as suffering from leukaemia.

The Ministry of Defence eventually accepted that the radiation had caused the cancer and that they were liable to pay the plaintiff damages, and after a court hearing the sum of around £170,000 was paid out to Mr Molinari.

The Armley environmental asbestos case (1996)

The two plaintiffs in this action were women who lived in the Armley district, in Leeds. They both lived close to the asbestos factory run by J.W. Roberts Limited, which is closely linked to Turner and Newell Ltd, one of the world's major producers of asbestos.

The plaintiffs contracted the fatal cancer mesothelioma, the only known cause of which is exposure to asbestos. The two women had played, as children, in the immediate vicinity of the factory and it was alleged that they were exposed to asbestos as a result of the plant operators allowing asbestos dust to disperse into the local area. Evidence was given to the extent that the asbestos pollution was very extensive indeed, with accounts of the streets being covered in large quantities of asbestos dust and fibres.

The court held that J.W. Roberts Limited should have been aware of the risk to human health of polluting the Armley neighbourhood with asbestos dust, that the company had breached its duty of care and that it was the asbestos from the plant that caused the cancers in the two women. Damages were therefore awarded.

This is an important case because it concerns a truly environmental impact of a factory process on its neighbours, and this case may have very important implications for industry. However, in considering the ultimate significance of this case it must be appreciated that the physical amount and quantity of the pollution was extreme. Also the scientific proof of the health effects of asbestos have now been studied for many years, early reports of concern being expressed by HM Inspector of Factories as early as 1890. Thus a cynic may say that it has taken 100 years for the asbestos industry to be brought to book!

The difficulty with many other environmental impact claims is the relatively small amounts of poisonous substances usually involved (such as dioxin in soil and herbage) and their implications for risk to human health. As has been seen with cases like the leukaemia victims living around the Sellafield plant, proving the causative link for other pollutants can be very difficult.

Benlate/Benomyl (1993)

There is great controversy concerning the fungicide Benomyl, now in the spotlight because it is alleged to have caused children to be born without eyes. This issue has been widely reported as a campaign story in the *Observer* and *News of the World* newspapers. American lawyers have taken one successful case to court in Florida, where they sued the manufacturers, DuPont, on behalf of an American child who was born without eyes following a Benomyl tractor spraying incident. The jury awarded the child $4 million.

The US lawyers will file claims on behalf of a number of other similarly afflicted claimants, including some from Britain. In the US, a jury hears a personal injury legal action whereas in the UK it is by judge alone. In the US higher levels of damages are awarded. There are also some advantages to a claimant in the US court procedure, with differences even between states. However, there are often lengthy arguments of 'forum' raised at the outset of a case, about the appropriate venue where a case should be heard, and increasingly the US courts, which have over the past few years become burdened with foreign claimants, are reacting against this and are refusing to allow foreign claimants to bring actions, including group actions.

Organophosphate pesticides and sheep-dips

Organophosphates (OPs) and injury caused by their use in sheep-dips have resulted in a great deal of media comment. OPs are inherently dangerous. They are highly toxic. OPs vary in levels of toxicity. A potent OP is 'Sarin', the nerve gas used in the attack on commuters in the Tokyo underground, and in the Iraq–Iran war. It is useful to consider the background of sheep-dip litigation. Matters came to a head with allegations by many farmers of serious ill health following the dipping of sheep with OP pesticides. The intention of dipping sheep is to eradicate animal parasites that are injurious to sheep. The products used by farmers were and continue to be licensed as veterinary medicines (rather than as pesticides by the Advisory Committee on Pesticides) by government upon recommendation from the Veterinary Medicines Directorate (VMD).

A number of claims are now in progress, mainly alleging that OPs have caused quasi-neurological or neurobehavioural illness. Admissions by ministers of possible injuries caused by OPs used in the Gulf War has resulted in further research.

A lead sheep-dip case is that of Mr Gary Coomber, a Kent farmer, who has issued proceedings for personal injuries as a result of his exposure to two sheep-dip products, Coopers' dip and a Youngs' Animal Health dip. Mr Coomber suffers from heart disease. This case is brought against the manufacturers of the sheep-dip products.

Yates v. *Rentokil (1992)*

Claims for personal injuries can arise in rare or idiosyncratic circumstances. For example, Barry Yates sued Rentokil after contracting soft tissue sarcoma, an extremely rare cancer, following exposure to pentachlorophenol. It was said that he used 22,000 litres in confined spaces treating for woodworm. His potential occupational exposure was, therefore, extremely high. Pentachlorophenol may contain dioxin impurities. An out-of-court settlement of £90,000 and substantial costs resulted after prolonged litigation. The incidence of this disease is about one in 6 million of the population.

CONCLUSION

How effective is personal injury legal action in achieving change?

One indirect consequence is that litigation may spur on research that has often been sadly lacking. It will spotlight media and public concern. While the system of compensation relies upon proof of fault before awarding compensation, it is the only way forward. However, there are suggestions for improvement. The most radical solution (and most unlikely) is compensation on a no-fault basis. Such a campaign for the victims of chemical poisoning was launched in the House of Commons in March 1994 with support by the Labour Party for its introduction.

At the present time, individuals who are damaged by chemicals and pollution may well find themselves without redress. Unless the standard of scientific proof is met, and proof is established to legal satisfaction, individuals will not be compensated. As a result, entitlement to receive compensation is a lottery. Without fault there is simply no case. Claimants should therefore be aware of the limitations of legal action.

Often, important issues and cases are ended abruptly when the polluting company 'pays off' the claimant (through payment into court). There is no decision on liability. In important and exceptional cases there should be a trial of that issue (at public expense) to resolve the same in the interests of future claimants. Whatever the future, there must be an improvement in a system widely considered by the public as being there to enrich lawyers and encourage paper-shuffling.

13 Loss and Damage

People's primary concern about environmental pollution is without question its potential effect on health. However, another inevitable and probably more widespread consequence of pollution is the devaluation of those properties situated in its vicinity.

Any property that is significantly exposed to substances associated with adverse health will be devalued (blighted). This is especially so in a buyers' market. After all, what sane buyer would choose to buy such a property in preference to a similar one in an unpolluted area?

Taking an extreme example, if a particular property is known to have been built on a former site of low-level radioactive waste, it is very unlikely that anyone would want to buy it. The fact that experts have advised that the site is no longer emitting significant levels of radioactivity would make little difference. This is because, despite the expert advice, buyers would probably be disinterested because of the risk to their health (no matter how small) and the knowledge that future buyers of the property might be put off by its location for the same reason. Certainly, a building society or other institution that was approached to lend money for a mortgage on such a property would be unenthusiastic unless the purchase price was substantially lower than a similar property elsewhere.

In this chapter, the extent to which legal action can be used to obtain compensation for property devaluation is considered. As will be explained, the fact that a property has been blighted will give rise to a valid claim in law only in very particular situations.

CAUSES OF DEVALUATION

The following are examples of environmental pollution that may cause devaluation of a property. This is not intended to be an exhaustive list but simply an illustration of the type of problems that commonly arise.

Toxic waste

A typical example of this would be one of the many properties around the country built on former gasworks sites (of which there are hundreds in the UK). The soil on these sites may well contain a cocktail of hazardous chemicals that pose a variety of health risks ranging from skin rashes to cancer.

The Winsor Park Estate in East London, comprising several hundred homes, was completed in around 1990 on the Beckton Gasworks site. Although measures appear to have been taken to 'cap' the site to prevent upward migration of hazardous chemicals to the surface of the soil, residents have been extremely concerned, especially about possible risks to their children's health. Concern was heightened when it was discovered that the water pipes leading into individual properties on the estate were made of a type of plastic that could be permeated by the very type of hazardous chemicals present on this type of gasworks site.

Methane

Decomposition of biodegradable waste beneath property produces methane gas, which may accumulate giving rise to a risk of explosion, and there have already been a few incidents of properties exploding or catching fire as a result of methane build-up both in the US and in England (for example, Loscoe, in Derbyshire, in 1986).

The Ebourne Estate in Kenilworth, Warwickshire, comprises some 90 properties built by the Barratt group in 1982, on an old landfill site. A mixture of domestic and industrial refuse had been deposited on this site for many years before the construction of the estate. In 1990, the local authority discovered elevated methane levels on the estate. One house caught fire and as a result there was a dramatic devaluation in property prices. Residents of houses on the part of the estate that is believed to overlie the old waste tip have been forced by the council to install mechanical methane-extracting equipment on their properties. This equipment requires continual maintenance.

Two of the residents sued for compensation against their surveyor for failing to warn them of the potential risk. Similar actions are likely to arise in the future.

Local industry

Emissions of noxious/harmful substances from the local industry range from unpleasant fumes or dust to non-perceptible carcinogens.

In January 1993, the television documentary *Storyline* alleged that a high level of cancers had occurred among the workforce at a pesticide plant and in the residents of a street running along its length. Pesticides have been linked with cancer. As well as causing widespread panic, properties in the street were immediately devalued. However, scientific studies carried out by the HSE subsequently failed to show that there was in fact any excess of cancers in either the workforce or the residents. Legal actions were nevertheless pursued by the residents, not for cancer, but for distress and inconvenience caused by emissions of unpleasant fumes from the plant, which in one instance had persisted over several weeks.

A specific category of pollutants that have been of considerable concern over the last few years are dioxins, furans and polychlorinated biphenyls (PCBs), a substantial proportion of which are emitted from chemical and clinical waste incinerators. Dioxins and PCBs accumulate in body fat. Several epidemiological studies have linked these chemicals to birth defects, miscarriages and cancer. The International Agency for Research on Cancer issued a press release in February 1997 confirming that dioxins are human carcinogens at high doses, increasing the risk of cancer by a factor of 1.4.

In the UK, concern about these man-made chemicals has grown. Throughout the UK, local residents have been involved in active campaigns against the siting of incinerators in their area, with campaigns focusing on opposing the planning applications lodged by the prospective operator. While these campaigns are motivated by concerns about the health of the community, the effect of an incinerator on property prices is undoubtedly also a factor in the protests.

On 21 February 1996, Coalite Products Limited pleaded guilty at Leicester Crown Court to emission offences concerning dioxins, and were fined £150,000 and costs of £300,000. The offences arose from the operations of an incinerator in Derbyshire, where the milk from cows grazing in the area was banned from sale. A legal action against the company for damage to property and loss of income from milk sales by three farmers was successfully pursued.

A claim for property devaluation by a non-farming couple living in the same area in respect of the same contamination is currently, at the time of writing, being pursued in the High Court in London. The trial is due to take place in December 1997.

Another typical example of land contamination is petroleum leakage from defective storage tanks in petrol stations. This may result in contamination of soil and ground water requiring extensive clean-up works. It has been recognised as a major problem in the US but the extent of the problem in the UK is at present unknown. Given that ingredients of petrol, such as benzene, are known carcinogens, significant contamination will, unless removed satisfactorily, cause devaluation. A couple living in Staines (south of London) are, at the time of writing, involved in proceedings against Texaco following the flooding of their land by a neighbouring petrol station.

Noise

Noise levels high enough to affect quality of life (for example, disturbing sleep and the watching of television) will seriously lower the value of property (as compared with properties of a similar type that are unaffected by noise). However, noise could rarely give rise to a valid claim in law for property devaluation.

There are three reasons for this. First, as will be explained below, it is only where pollution is the result of an unlawful activity that a claim can be sustained. Therefore, the fact that a property, for example, next to Heathrow Airport is subjected to noise from low-flying aircraft will not give rise to a claim unless the noise exceeds permitted levels. Further, even where noise is attributable to a lawful activity (for example, a neighbouring factory operating at night) it may be possible to stop the problem by applying to the court for an injunction to stop night working. If such an application were successful this would eliminate the problem and no devaluation of property should result.

The third reason is one that applies to any claim for economic loss, as the law presently stands. Apart from certain exceptions, a claim for economic loss cannot be sustained unless some 'physical damage' has occurred. The legal meaning of this phrase will be discussed further below. Suffice it to say that noise does not constitute 'physical damage'.

Fly infestation

Fly infestation is not a problem that obviously springs to mind in the present context. However, there has been considerable publicity recently about fly infestation from egg-production plants. A typical plant housing 100,000 chickens may produce up to one tonne of chicken waste a day. This is obviously an ideal breeding ground for flies, which, if not properly controlled, are not only a real nuisance but also can create serious health risks. Again, as in the case of noise, the availability of injunction proceedings to reduce the flies to a reasonable (lawful) level means that a claim for property devaluation will rarely arise.

Radiation

Radioactive contamination has been a major public concern since the development of the atomic bomb during the Second World

War, although the perceived source of the risk has switched, in recent years, from man-made radiation, such as from nuclear reprocessing plants, to naturally occurring radiation, for example, that emitted from rock in the form of radon gas. The fear arises from the known association between radiation and cancer, particularly childhood leukaemia.

In 1989, Mr and Mrs Merlin, a couple living close to the Sellafield Nuclear Reprocessing Plant in Cumbria, pursued a claim against British Nuclear Fuels plc for property devaluation caused by elevated levels of radiation in their house dust. The action failed for two primary reasons. The judge ruled first that the level of contamination did not constitute 'physical' damage to the property and second that the levels of radioactivity did not present a 'significant' risk to health. The fact that to cause physical damage to the property the radioactivity would have had to have been at a level that would have killed all the occupants many times over did not, to the judge, seem an anomaly.

The Merlins' case is to be contrasted with the success in 1996 of Blue Circle Cement in holding the Ministry of Defence liable for radioactive contamination of land from the Nuclear Weapons Establishment in Aldermaston, Berkshire. It is of interest to note that the level of contamination in the company's property was many times lower than the level in the Merlins' house dust. The judge distinguished the Merlins' case by reference to the question of physical damage. The Judge held that a physical change in the property had occurred because Her Majesty's Inspectorate of Pollution had recommended clean-up of the land by removal of the radioactive material. No such recommendation was made with respect to the Merlins' house because, unlike Blue Circle's land, no authority was responsible for regulating it. Viewed cynically, Blue Circle's success may simply reflect differential treatment of commercial concerns and private individuals. Even if that were so, the new monitoring obligations (yet to come into effect) of local authorities under the Environment Act 1995 are likely to alter the position.

Radon gas occurs naturally and is emitted in significantly higher levels by rock such as granite. Thus radon levels in properties in regions such as Cornwall and Derbyshire have been found to have significantly higher levels than the background level generally in the UK. Radon is a source of alpha radiation which, though less penetrating than X-rays, is believed to deliver a dose of radiation at least twenty times higher when ingested or inhaled. There is a body of reputable scientists who argue that naturally occurring radon may well be a contributory factor in causing childhood leukaemia.

Radon levels are lower in well-ventilated properties, which means that the levels are increased in houses that have been double glazed, an ironical consequence of modern technology.

PREVENTATIVE MEASURES

It is obviously sensible to avoid buying a property that is likely to be affected by environmental pollution. There are various enquiries that can and should be made to try to identify potential at-risk properties. In March 1993, the UK government withdrew plans for a register of contaminated land, which would have made identification easier. However, in more recent years some local authorities have been maintaining their own informal registers. Under the 'Contaminated Land' provisions of the Environment Act 1995, local authorities will be obliged to carry out monitoring to ascertain affected areas. These provisions are likely to come into force in the near future. The Act was intended to bring into force the 'polluter pays' principle.

In any event, prospective purchasers should make detailed enquiries, including the following:

- Ensure that valuation surveyors specifically consider the issue of contaminated land in their reports.

- Ensure that conveyancing solicitors ask the local authority specific questions about contamination on or around the property before exchange of contracts. Solicitors will invariably undertake a search of the local authority registers before exchange of contracts. However, this will probably be done using a standard search form, which is unspecific about contaminated land.

- Make specific enquiries about releases of unpleasant or toxic fumes from any neighbouring industry.

- Ask for confirmation of whether any planning applications are pending for activities that might affect the land value at a later date. For example, have any applications to operate a toxic waste incinerator in the area been submitted? (See Chapter 7 for more information on the planning process.)

If the new house is being bought from a builder or developer, the prospective purchasers should carry out additional research:

- Seek information about the history of the site on which the property is built.

- Obtain a copy of the site survey. A survey is likely to have been undertaken by engineers, and should include an analysis of the soil composition.

- Obtain details of the builder's public liability insurance. It is, unfortunately, not uncommon for builders to go into liquidation. An owner therefore needs to be in a position to refer any future claim to the builder's insurers. Identifying insurers is a notoriously difficult task.

- Request information about the site from the National House Builders' Council (NHBC). Any new property will be covered by a certificate from the NHBC. This is an insurance policy covering structural defects in buildings. Whether the NHBC will pay out for the devaluation caused by the presence of contamination of land is at present unclear. Clearly the NHBC policy will not cover devaluation caused by antisocial neighbouring industry. However, before issuing the NHBC certificate, the NHBC (like any insurer) will have made various enquiries about the nature of the site. There is no reason why the NHBC cannot be asked for this information.

None of this is of any use to those already living on properties that have been devalued by pollution. Here the primary issue is compensation. Whether or not this is available depends in turn on consideration of legal liability. Different legal issues arise, depending on whether one is considering liability on the part of a private seller, building developer, polluting industry, local authority, surveyor or solicitors.

COMPENSATION CLAIMS

Unless it can be negotiated, obtaining compensation for devaluation is dependent on persuading a court that someone ought to be held legally responsible for the devaluation. Since the objective of the exercise is financial, it is obviously important to choose the target of any legal action (the defendant) carefully. The best defendant is the one who is most likely to be held responsible and who has the means to pay any award by the court. This is not necessarily the same as the one who is considered to be responsible from a moral perspective.

There are various considerations to be made in deciding which, if any, legal course to pursue. The main ones are the following.

Potential defendants
Who are the potential defendants and what is the nature and scope of their legal liability – that is, what factors will determine whether or not a particular defendant is liable and for what might they be liable? This is discussed in more detail below (page 330).

Link between pollution and devaluation
What evidence is available to prove the link between the pollution and the property devaluation? This is not always as straightforward as it may seem. For example, in the case referred to above involving the pesticide factory, property devaluation was also alleged. As well as the clear evidence of noxious emissions from the plant, the television documentary mentioned above had linked the plant with cancers in the area. The difficulty lay in trying to disentangle the extent to which the emissions rather than the cancer scare had caused the devaluation. Naturally, the company contended that any devaluation was solely due to the cancer scare.

Degree of devaluation

By how much has the property been devalued? Evidence will need to be obtained that demonstrates that the property is worth less than it would have been if the pollution had not occurred. The fact that prospective purchasers may have lost interest once they had become aware of the pollution is important evidence. So too is any supportable suggestion that properties in a different area unaffected by the pollution, which used to have a similar value to the property in question, are now selling for much higher prices.

Previous assurances

At the time the property was purchased were any assurances or advice given about pollution of the land that have turned out to be incorrect? In one case involving an estate in Sevenoaks, Kent, a builder/developer who was specifically asked whether any biodegradable material was present on the site responded confirming that this was not the case. A few years later, high levels of methane gas were discovered on the property. Advice given by the purchasers' surveyors should also be scrutinised. What did they say about the property? Was any mention made of the possible risks associated with the present pollution on the land or the fact that the land had been subject to environmental pollution?

Means to pay compensation

Of those potentially responsible for the pollution, who is likely to have the means to pay compensation for a property devaluation? There is no point bringing a claim against someone who does not have the means to pay. As mentioned above, it is not uncommon for building companies to go into liquidation. This may not matter provided their insurance details are known (and the policy covers the loss). It is now unfortunately not uncommon for claimants to have to invest considerable time and expense trying to track down the relevant insurers, often to no avail. This is one reason why it is important to obtain details of insurance at the time of purchase. In this context it is worth noting that, since about 1990, insurers in the UK have declined to issue policies covering contaminated land. Consequently a builder's insurance is likely will only to be of assistance in the present context in the case of properties constructed before 1990.

Time limits

Is it too late to bring a claim? The rules vary depending on whether a claim is brought under a contract, for negligence, nuisance or breach of statutory duty, or under the terms of a lease. The general rule is that claims for breach of contract, negligence and nuisance must be brought within six years. Thus a breach of contract claim by a buyer against a seller for failing to disclose the fact that land on which a property was built was polluted would need to be made within six years of the date of exchange of contracts. A claim for breach of contract against a surveyor for failing to identify a pollution hazard must be brought within six years from the date of the surveyor's report. A claim against a freeholder of the property – for example, for breach of the covenant of quiet enjoyment in a lease – must be brought within twelve years from the date of the breach, that is, when the pollution occurred.

These limitation deadlines can be complied with if the pollution is obvious. However, it is often the case that pollutants on land will remain dormant and unnoticeable for many years after the contamination occurred (by which time the limitation deadline may have expired). Fortunately, the Limitation Act 1980 allows that claims against surveyors and solicitors in negligence can also be brought within three years of the date when the claimant ought reasonably to have been aware of the damage.

TARGETS FOR LEGAL ACTION

It is essential to bear in mind that, as has already been mentioned, there is no entitlement to compensation for devaluation that arises inevitably from an activity that has been expressly authorised by statute or the common law. If, for example, planning permission is granted for the expansion of a road, it is inevitable that houses next to it will be subjected to greater traffic pollution and noise and may be devalued as a result. The house owners will have no claim under the law for this devaluation as it will be assumed that the loss was authorised by Parliament.

Having said that, in such a situation a statutory compensation scheme may be set up. If the road scheme results in an increase in the levels of noise in neighbouring properties, the insulation of the windows may be made by the Highways Agency. If, on the other hand, the road is poorly constructed so that local residents wake up whenever night traffic passes over bumps on the road, then a claim in negligence for compensation might well arise (against whoever was responsible for constructing the road) in the event that this deters prospective purchasers from buying the property.

In general, the same principles apply in relation to the responsibility for devaluation of contaminated land as applies to any other type of damage, such as defects in the structure of a building. There is a variety of potential targets for legal action, which broadly fall into the categories set out below.

Sellers

The fundamental and often quoted maxim, *'caveat emptor'* ('buyer beware') is generally applicable. This effectively places the burden on a purchaser to make enquiries that might reveal any problems. There are, however, the following exceptions to the rule.

Misrepresentations

Where a purchaser asks a specific question and is misled by the seller, then, provided the purchaser can demonstrate that this resulted in financial damage, the seller will be contractually liable. An example of this situation is the Sevenoaks case referred to above (page 329).

Latent defects

If a defect in a property is known to the seller but would be difficult for the buyer to discover, the seller may be held liable for loss resulting from such a 'latent' defect.

Builders

Unless one is dealing with a large established building company, there is always the worry that, even if a building company can be held legally liable, it will be unable to pay – for example, if the company has been liquidated. For that reason, as has already been emphasised, the insurance details should be ascertained at the outset. If the insurer can be identified, the next stage is to check that the policy covers the loss suffered and that the insurer does not have grounds for avoiding liability under the policy.

While insolvency is of less concern with respect to large builders, the downside with them is that over the last ten years they seem to have adopted a practice of purchasing large areas of contaminated land, which they then develop as and when there is a demand. This phenomenon was highlighted in a recent *Dispatches* documentary.

Where devaluation is caused by the presence of hazardous material on the ground beneath a property, this clearly does not fall within the definition of being a defect in the fabric of the building. Nevertheless, there are four main grounds upon which claims can be brought against builders.

Breach of NHBC Agreement

In the case of a newer property (less than ten years old) where the original purchaser is the owner, there will be a direct contractual relationship between the builder and the purchaser. This would be governed by the NHBC Agreement. The standard form of Agreement contains a warranty, on the part of the builder, that the property has been 'built' in an efficient and workmanlike manner, using the proper materials so as to be 'fit for habitation'. The precise meaning of this clause has not yet been tested in the courts. In particular, it is unclear whether a court would accept that the presence of contamination in the land upon which the property had been built would be regarded as proof that the property itself had not been built in an 'efficient and workmanlike manner using the proper materials'. On the other hand, there might be a better prospect of proving that there had been a breach of the warranty that the dwelling has been built so as to be 'fit for habitation'.

Although the NHBC agreement expressly states that subsequent purchasers can take the benefit of the original agreement, a subsequent purchaser can only enforce the agreement against the builder, if the benefit of it has been assigned to each new owner of the property.

Defective Premises Act 1972

The Defective Premises Act 1972 applies where there is no NHBC agreement. This Act imposes obligations on the builder that are similar to those contained in an NHBC agreement. NHBC cover is to be preferred over use of this law as only the NHBC agreement protects a purchaser in the event of an insolvent builder.

The point is that if the builder was a company in liquidation it could prove difficult to trace the insurer. Even if the insurer and a policy are identified, the question that will arise is whether the policy covers the loss. The fact that finding the insurer is not the be-all and end-all was demonstrated in a recent decision of the Court of Appeal. The owners of a property that had exploded as a result of the accumulation of landfill gas beneath it from a neighbouring waste tip sued Commercial Union, the insurer of the waste-disposal company. The policy covered claims arising from the 'method of disposal' of the waste. The court held that it was not the 'method' but the location of the waste that had caused the damage, and consequently the policy did not cover the loss. Although the case did not involve a builder, the same insurance pitfalls are generally applicable.

Negligence

In order to succeed in negligence, a plaintiff must show that the defendant should have realised that specified acts (or omissions) might cause injury but yet took no steps to prevent the injuries. Negligence actions are a useful tool where 'physical damage' has occurred; however, property devaluation claims arising from pollution may be regarded as purely financial involving no actual physical damage (see section on noise, Chapter 14). When it comes to claims for pure financial loss – for example, a claim against a doctor for loss of earnings on the basis that the plaintiff gave up a lucrative job on the strength of the 'negligent' advice that he had only a short time to live – the general consensus is that such a claim (in negligence) would fail because no physical damage has been caused. On that basis, the test in the present context is whether contamination of land constitutes 'physical' damage. Liability of builders has been the subject of extensive case law, the upshot of which is that a builder is unlikely to be held liable (in negligence) for devaluation, even if there is physical damage to property.

Thus in the case of a property constructed on a landfill site that is affected by high methane levels, a builder is unlikely to be held liable, even though this risk was known in the building trade as early as 1979.

The rationale behind these decisions appears to be public policy: that is, the court's concern that by granting the owner of a property the right to sue a builder for negligence, the 'floodgates' would be opened.

It should be noted that these difficulties in holding a builder liable in negligence will not enable the builder to escape if there is evidence of misrepresentation or fraud.

Breach of covenant of 'quiet enjoyment'

Where an occupant of a property is a lessee (the tenant who takes on the lease), it may be possible to claim compensation for breach of the covenant of 'quiet enjoyment'. This is a standard covenant given by a freeholder to a lessee. For example, liability has been established where the occupation of leasehold premises was made intolerable because of noise, dust and dirt.

Local authorities

Victims of property devaluation will invariably have an axe to grind with their local authority. This may be because of a feeling that the council should not have licensed the dumping of waste on the land and/or granted planning permission for the houses on it. Furthermore, once contamination is identified, action taken by the council is often responsible for inflaming the situation. For example, in the Kenilworth case described above (page 321), the council effectively warned off prospective purchasers of property by placing a caution on the Register of Title of individual properties that specifically referred to the landfill gas investigation. Subsequently, the council threatened to take statutory nuisance proceedings against residents who were unhappy about installing methane-extracting equipment on their properties. There are many similar examples. In such situations the council creates a 'triple whammy': licensing the dumping of waste; granting planning permission; threatening action for a problem that it should arguably have prevented in the first place. In these circumstances it is understandable that the victims regard the council as being morally to blame. Unfortunately, moral blame does not equate with legal responsibility.

The only feasible basis for a claim against a local authority would be in negligence. However, again due to public policy considerations, the courts have consistently refused to impose a duty on local authorities except where one is expressly imposed by statute. In particular, the courts will be reluctant to impose liability where the problem has arisen out of a local authority's general responsibility for public welfare. Both the licensing of the dumping of waste on land and the grant of planning permission for development of the site would appear to fall into that (general) category. This contrasts with action taken by a council, with reference to specific individuals. Having said that, if, prior to purchase of a property, a local authority has been asked specific questions about pollution of the property, which it answers carelessly, and loss arises, then an action in negligence may well be sustainable.

Both builders and local authorities will, therefore, generally be protected from liability by public policy considerations. If this were not the case, a claim against either of them might succeed if it could be shown that they ought to have foreseen the risk. The general view is that risks associated with the build-up of landfill gas have been known at least from the early 1980s. Indeed, a building research establishment paper in 1979 expressly drew attention to these risks and stated that sites containing decomposable material – for example, refuse dumps – 'were best avoided for building purposes'. Thereafter, the issue was covered by the 1985 Building Regulations.

Dumpers of waste

Where waste has been dumped in contravention of a Control of Pollution Act 1974 licence, an offence is committed and liability will arise under section 88 of that Act. Otherwise liability may arise in the common law. This has been clarified by the 1994 House of Lords decision in *Cambridge Water Company* v. *Eastern Counties Leather plc* (see Chapter 12, page 294, and Chapter 14, page 352). The rule is, essentially, that anyone who causes damage to a neighbour's land will be liable for the consequences, provided the damage was reasonably foreseeable. This may be sufficient where the damage is contemporaneous with the contamination – for example, emissions from a factory polluting a residential area. It is not sufficient, however, where the damage takes a long time to manifest itself, as was the case in the Cambridge Water action, or where at the time of the dumping of waste there was no reason to believe that properties would later be erected on the site. Furthermore, a court would be highly disinclined to impose liability in favour of unspecified future owners whose loss did not arise until some time in the future.

Surveyors

While perhaps not springing to mind as being the most morally culpable target, surveyors may prove to be the most practical targets for legal action in contaminated land cases. What is being considered here is the responsibility of surveyors (either valuation or structural surveyors) who undertake pre-contract surveys upon which purchasers and lenders rely.

There is clear authority to the effect that a building society valuer (instructed by the lender) owes a duty of care to a purchaser (see Chapter 12). A surveyor is not relieved of this duty simply because he/she undertook the cheaper (valuation) version of a survey. Consequently, a surveyor can be held liable in negligence for property devaluation, even where there has been no physical damage. This is in important contrast to the position of builders and local authorities.

Although surveyors owe a duty of care to prospective purchasers, it is still necessary to prove that the surveyor breached the duty of care and that the breach caused the loss. The examination of these issues will depend on the nature of the contamination. Was damage resulting from the contamination of the site in question foreseeable at the time of the survey? If the surveyor had warned the purchaser of the risk, would it have made a difference, or would the purchaser have bought the property at the same price in any event? If the risk was not reasonably foreseeable or the purchaser would have gone ahead even if a warning had been given, then the claim will fail.

For example, with respect to methane gas, the danger of landfill sites was drawn to the attention of surveyors at the beginning of 1988 in an article in *Chartered Surveyors Weekly*. From then on, the profession should arguably have been on notice of this risk and anyone who was not warned may have a claim in negligence. Prior to 1988 the position is unclear, although some surveyors were warning of this particular hazard from as early as July 1985. However, the question is whether a 'reasonably competent surveyor' would have given a warning.

Another example, which is likely to crop up in the future, is the presence of lead paint, particularly in old properties. Before the 1960s, gloss and primer paint contained a high percentage of lead pigment. Removal of paint (by heating or sanding) prior to redecoration liberates the lead into vapour or dust, which can cause injury to the inhabitants of the property (including the unborn child, if redecoration is undertaken during pregnancy). Even though properties repainted since the 1960s will have a top coat with a lower lead content that previously, the old painted surface will pose a health risk if rubbed down with a sander or vaporised with a heat gun. Ideally, this hazard should be removed before the purchasers move in with the purchase price being negotiated down to cover the cost of eliminating the hazard.

A surveyor who fails to warn a purchaser of this risk could quite conceivably be held responsible for the cost of removal, or the making safe, of the lead paint.

One final example is in relation to the alleged link between electromagnetic fields and cancers (particularly childhood leukaemia). The evidence that there is a link has been gathering since 1979, but it was not until 1996 that the Royal Institute of Chartered Surveyors inserted a suggested clause in their so called 'red book' (which advises their members what points to cover within their surveys). The clause, informing the purchaser of the presence of a source of higher than normal levels of electromagnetic fields, goes on to make the point that the controversy surrounding the link could impact on the value of the property. This is all said, rightly, with the surveyors specifically setting out that the rights and wrongs of the scientific argument cannot be matters for them.

Conveyancing solicitors

Solicitors engaged on house purchases have a clear duty to protect their clients. The liability on the part of the solicitor is dependent on the nature of the pollution involved. However, it is likely to be harder to claim compensation from a solicitor than from a surveyor. This is because the nature of the risks involved are generally more foreseeable to surveyors than they are to solicitors.

THE ENVIRONMENT ACT 1995

The prospective coming into force of the contaminated land provisions of the Environmental Protection Act 1990 via the Environment Act 1995 have already been referred to. It is inevitable that a property identified by a local authority as being contaminated will be devalued. This in turn will undoubtedly result in compensation claims, provided the victim of the devaluation can locate a worthwhile target to sue. Assuming the rationale of the Blue Circle judgment remains intact, then the fact that the local authority has stipulated the carrying out of remedial works to remove the contamination would be sufficient proof of 'physical' damage.

This, in turn, will strengthen claims, particularly those against builders, where the main obstacle to legal redress is currently the inability to prove 'physical' damage. It remains to be seen whether this predicted increase in claims will materialise. A lot may depend on whether funding is available to pay for the legal costs of such an action. In that regard, it is possible that building societies and banks who have lent money, by way of mortgage, for the purchase of affected properties will become involved in the funding of actions, possibly even making their own claims.

CONCLUSION

Actions for property damage have not been that common or successful to date. This is a key area for development if people's rights are to be protected by the courts. It is to be hoped that individuals affected, notwithstanding the difficulties, will consider looking to the court process to obtain redress. Only in this way will the power of the polluters be really undermined.

14 Nuisance

It is common to describe pollution as a 'nuisance'. However, few people appreciate that there is a legal remedy that also uses that title. What does a legal action for nuisance involve?

The following scenarios are examples of where a nuisance action might resolve a particular problem.

- A person becomes depressed through loss of sleep as a result of a neighbour continually playing music into the early hours.

- A family is unable to use the garden because of odours from a nearby factory.

- People are forced to move out of their home because of an epidemic of flies from a farm where hygiene standards are poor.

- Methane gas from a landfill site has leaked into the foundations of a housing estate, creating a danger of explosion and a consequent reduction in property values.

In all these situations, the people affected are potential plaintiffs in legal actions against those causing the nuisance.

PRIVATE AND PUBLIC NUISANCE

In the past, the courts have divided this type of action into the two categories, called 'private' and 'public' nuisance claims.

An individual who has an interest in land, being either a property owner or a tenant, and who is suffering a nuisance, is able to claim under the law of private nuisance. Private nuisance is a remedy that has evolved through the common law to protect the rights of those who own or have an interest in land.

Alternatively, people who are neither owners nor tenants, and who have no particular legal connection with property, are able to bring an action in public nuisance if they can show that they are particularly affected by an activity that is causing a nuisance to a group of people. The requirement to show that an individual is more seriously affected by a nuisance than other members of the group may pose technical difficulties to bringing an action under this head of claim.

In 1995, the Court of Appeal considered the cases of *Hunter and Others* v. *Canary Wharf Limited* and *Hunter and Others* v. *London Docklands Development Corporation*, which were actions by residents of the London Docklands area who were claiming that the dust caused by the building of the site was a nuisance, as was the blocking of their TV signals by the erection of Canary Wharf. In doing so, the court was asked to rule on the principle of whether or not it was a necessary prerequisite to have a legal interest in property in order to claim in private nuisance. The Court of Appeal's decision was that as long as the claimants live in the property affected as their home then it is not necessary to show they also have a legal interest in property to bring a claim in private nuisance. This may have exploded the distinction between the two types of legal remedies available in the law of nuisance (public and private).

If, therefore, people, including children, who are not the property owners or tenants can show they live somewhere as a home and they are experiencing a nuisance, they are able to bring a claim in private nuisance rather than rely on the more outdated remedy of public nuisance. On the one hand, the decision removes what was essentially a discriminatory requirement for those who wish to bring a claim in private nuisance but could not because they had no legal interest in land, and at the same time may render the law of public nuisance redundant.

That decision was appealed to the House of Lords who gave judgment on the issue in 1997. The Lords (by a majority of 4 to 1) ruled that the law would return to the traditional notion of private nuisance being an action in respect of land. Any inconvenience or annoyance caused to people living within the property would only be compensated as a loss to the amenity value of the property. Accordingly, there could only be one claim per household and that claim could only be brought by the freeholder, leaseholder or tenant. If the property is jointly owned or rented by two or more individuals, the damages must be split between them.

Lords Hoffman and Hope went further, ruling that the calculation of damages for loss of amenity in transient nuisances is based on pure economic values rather than by reference to comparable personal injury cases. The damages would not be increased because more people lived in the property, were present for longer periods during the day, or were more seriously affected.

The damages would be: the value of the right to live in the property without nuisance minus the value of the right to live there with the nuisance multiplied by the time for which the nuisance affects the property. It was envisaged that these values could be calculated by estate agents. Accordingly, the larger the size and value of the property, the greater the loss. Mansions with occasional single occupations, for example, would thus be awarded larger damages than family houses.

However, for the purposes of this chapter, no distinction will be made between private and public nuisance, but the general principles of what is involved in bringing a nuisance action will be considered.

NUISANCE GENERALLY

A nuisance action will usually involve a claim for compensation for damage to property and/or loss of enjoyment of property, and may include a claim for an injunction to prevent the nuisance occurring in the future. The proceedings can be brought by an individual or a group of individuals where many people are affected by the same activity. Legal aid is available to bring a claim for nuisance if the plaintiff is able to satisfy the means, merits and reasonable tests (see Chapter 3, pages 40–8).

However, unlike the law of negligence, which has been developed and refined in the last century, the law of nuisance has been used infrequently aside from neighbour disputes. It is only since the increased level of public awareness about the importance of protecting the environment generally, and a desire to avoid unnecessary levels of environmental pollution, that the law of nuisance has been revived by lawyers as the most appropriate legal remedy available in the common law to protect those who are subjected to unreasonable levels of environmental pollution.

Whereas the law of nuisance was previously considered useful only for neighbour disputes and was given scant consideration by lawyers, it is now regarded as a useful remedy that can be used to challenge factories that create pollution, or industry building works, the activities of which may blight whole neighbourhoods.

As a result of its infrequent use over the last century, the ambit of the law of nuisance is not clearly defined and is likely to be the subject of development through the appeals system in the coming years as plaintiffs in environmental actions rely increasingly on it to protect themselves.

Nuisance also has an advantage over negligence in that it is possible to apply for an injunction to prevent a recurrence of the nuisance, which is not possible in negligence. Furthermore, the level of fault that the plaintiff must establish in a nuisance action is less than the one that must be established in a negligence action.

Definition of a nuisance

In legal terms, a nuisance is defined as the 'unlawful interference with a person's use of or enjoyment of land'. The definition makes it clear that the law of nuisance has at its heart the protection of the use or enjoyment of property. Generally speaking there are three categories of nuisance:

- encroachment onto the plaintiff's land, where the nuisance closely resembles a trespass – for example, where a gutter hangs over and drips into the neighbour's property;

- physical damage to the plaintiff's land;

- interference with the plaintiff's use and enjoyment of land.

The first two claims are the more common ones. Generally speaking, it is easier to mount a successful nuisance claim where an individual can show that there has been physical damage to his/her land or property than to win a case where the individual is claiming that his/her enjoyment of the property has been affected by activities of his/her neighbours or a polluter.

Actual damage to person or property

Cases where there is physical damage include damage to the premises themselves, to plants and trees in the garden, to washing hanging out on the line, to curtains and to car paintwork. For an action to be taken the damage must be shown to be real, to be significant and to have been caused by the nuisance that is the subject of the complaint.

Many of the real or 'actual' damage cases involve claims of damage from airborne pollution from industrial processes. In the case of *St Helens Smelting Company* v. *Tipping* (see page 349 below), the plaintiffs complained about the effects of copper smelting. Other cases involve brick dust from brickworks, emissions from coking plants and gasworks, and smuts from oil-distribution depots that have caused damage to curtains and car paintwork.

In these types of case the plaintiffs will succeed in their claims for damages notwithstanding that the defendants – the people operating these processes – often put forward arguments to the general effect that they were using the best methods available to reduce or eliminate the level of pollution and that they were fulfilling an important function within the economy of the community by supplying goods and services and employing people.

This means that individuals who can show that they are suffering actual damage from a process have a very strong claim for nuisance even if the defendants can show that they are doing their very best to reduce the level of nuisance to the lowest level possible. Different considerations will apply if the operators can show that they are undertaking the process by virtue of a 'statutory' authority, but it is rare for industrial processes to be operated on this basis.

Sensibility claims

Sensibility cases, which relate to complaints of interference with comfort and convenience, are likely to arise from offensive smells, high dust levels, and from noise or vibration but where there is no physical harm. However, the types of disturbance that can give rise to an actionable nuisance are not closed. As long as the plaintiffs can show that their comfort or convenience has been materially disrupted so as to affect the reasonable enjoyment of the property, then an action for nuisance may be available to prevent that nuisance.

In the London Docklands case described above (page 341), the judge at first instance held that interference with television reception was capable of constituting an actionable nuisance. In his judgment, he stated that 'watching television plays an important part in an ordinary householder's enjoyment of his property, ... whether he watched for the purpose of recreation or entertainment or to receive news or educational programmes. Interference with television reception thus interferes with a householder in the comfortable and convenient enjoyment of his land.'

The Court of Appeal, reversing the judge's decision, held that television interference was not actionable at law. However, it is difficult to understand the reasoning behind this decision when television plays such an important part in people's lives.

The nineteenth-century case of *Walter* v. *Selfe* set out what individuals needs to establish if they are to succeed in a claim for loss of enjoyment of property. The Vice-Chancellor, Sir J.L. Knight Bruce, stated that they have to have suffered 'inconvenience materially interfering with the ordinary physical comfort of human existence, not merely according to the elegant or dainty modes and habits of living, but according to plain and sober and simple notions and habits'.

What this means in today's language is that the court will not allow hypersensitive plaintiffs to impose excessive restraints on their neighbours and that the court will apply a common-sense approach to what constitutes an infringement of the use and enjoyment of property and what does not. Thus, loud music at three o'clock in the morning is likely to be regarded as an actionable nuisance unless it is a one-off occurrence, whereas disturbance from DIY activities even very early on Saturday and Sunday mornings may not be.

Similarly, odours that are offensive to the extent that a family is unable to use its garden frequently is very likely to be held to be a nuisance, but this is unlikely to be the case for intermittent and occasional wafts of odour. However, a one-off odour that makes everyone in the family sick is very likely to be a nuisance because of the gravity of the effects. The decision is based on a question of degree and requires the court to weigh up the competing interests of the plaintiffs' right to use and enjoy their property and the defendants' right to conduct operations free from interference.

The Canadian case of *Devon Lumber Co. Limited* v. *McNeill and Others* (1988) in the New Brunswick Court of Appeal provides a good example of the court's approach to the 'hypersensitive' plaintiff. In that case the plaintiff suffered from a pre-existing allergic reaction to dust. The defendant ran a cedar mill which created fine dust that exacerbated the plaintiff's allergy. The court held that the plaintiff was not entitled to damages in relation to her allergy although she was entitled to damages for the general level of nuisance that the cedar mill caused. The exacerbation of her pre-existing condition was not recoverable using the yardstick of a normal person of ordinary habits and sensitivity as the measure as to what was reasonably recoverable and what was not. Although this and other commonwealth decisions are not directly applicable in the English law system, they can be of persuasive value, particularly when made by the higher courts of commonwealth jurisdictions.

It is useful to compare here the principles that the law of negligence would apply to a similar situation. In negligence, the 'eggshell skull' rule applies, so that the defendant must take the plaintiff as he/she finds him/her. If it had been possible in the Canadian case mentioned above to establish that Devon Lumber were *negligent* as to the amount of dust that they were creating in their operations, then the plaintiff would have been able to recover damages for the fact that her pre-existing allergy had been exacerbated by the operations of the mill.

How to prove the existence of a nuisance

It is very important for plaintiffs who claim they are suffering a nuisance to collect evidence that shows that the nuisance is occurring. For example, a diary of how the nuisance is affecting their daily life is probably the best evidence to put before a court. So in the case of a noise nuisance, the plaintiffs should keep diaries of when they are affected by the noise – for example, setting out the dates and times and what they are unable to do because of the nuisance or what they have to do to avoid it. Audio and video recordings can also be useful. Residents of Pentre Meilor near Wrexham in Wales, for example, sent video footage to HMIP of the Deeside Aluminium factory in their village emitting black smoke. HMIP subsequently prosecuted the owners of the factory, who were fined £20,000 and forced to install pollution-control equipment (see Harding 1997).

Evidence to the effect that it is necessary to move rooms in order to sleep, or that it is impossible to have a conversation because of noise levels, will be particularly compelling. Similarly, in an odour nuisance case, it is important for the person affected to describe what steps have to be taken to eliminate the odour. If, for example, windows have to be closed on a sunny day, thus making the house stifling hot, or a family is unable to use their garden or has to abandon a barbecue or an outdoor children's party, this will be compelling evidence that a person's use and enjoyment of his/her home is being adversely affected by the activities of his/her neighbour.

In addition, evidence of complaints to the local authority's environmental health department or the Environment Agency will also go some way to showing that the nuisance has occurred and that it is being treated seriously by the plaintiff. It is, therefore, important to keep a detailed record of the problems and complaints that have been made to the enforcement agencies as well as to the person causing the nuisance.

How will the court decide whether a nuisance has occurred?

A court will generally adopt a common-sense approach in deciding whether a nuisance is occurring. Essentially, the court's job is to undertake a balancing act between the activities of the two parties concerned. They will, therefore, weigh the activities of the defendant against the extent of the plaintiff's complaint and the way in which the alleged nuisance is affecting the ordinary conduct of the plaintiff's life. In undertaking this balancing act, the court will take a number of matters into account in approaching the central question to be determined on the facts of each case, namely, whether or not the defendant is using the property reasonably. Those factors most commonly given weight by the court are:

- the excessiveness of the conduct;

- defendant's malice;

- the character of the neighbourhood;

- continuous or repetitive incidents so compared with isolated incidents and the time at which the nuisance occurs;

- whether the best practical means have been used to remove the nuisance;

- the social value of the defendant's operation;

- whether the damage was foreseeable.

The excessiveness of the conduct
This rule speaks for itself and is fundamental in carrying out the balancing exercise. For example, one barking dog may be reasonable, 20 may well not be. The case of *Farrer* v. *Nelson* (1885) illustrates the point. In that case, land was used to rear pheasants which then became overstocked and caused a nuisance. It was not the conduct itself that caused the nuisance, but its excessiveness.

Defendant's malice

Motive may have legal significance and malice may render an otherwise innocent act an actionable nuisance. In the case of *Christie* v. *Davey* (1893) the plaintiff and defendant were neighbours. The plaintiff taught the piano and, although the noise was not excessive, the defendant could hear it. The defendant became irritated to the extent that every time there was a lesson he would bang on the wall. This was held to be a nuisance and the malicious motive was taken into account.

In the case of *Hollywood Silver Fox Farm* v. *Emmett* (1936), the plaintiff was a fox farmer and the defendant was his neighbour who held a grudge against him. During the mating season, the defendant let off a shotgun next to the premises, which scared the breeding foxes. It was held that, if the defendant had merely been shooting on his own land, this would not have been a nuisance, but as it was a malicious act it was actionable.

The character of the neighbourhood

This consideration applies only to personal comfort or sensibility cases, not where there is actual damage to property. An example is the case of *St Helens Smelting Company* v. *Tipping* (1865). Mr Tipping bought a property close to a smelting works and acid smuts damaged the trees on his land. The defendants said that he had to accept this as part of the locality of the neighbourhood. The court held that this was not a defence against actual damage. The court may have ruled differently if this had been a 'sensibility' case and had there been no damage to property or person.

The most famous case here is the nineteenth-century case of *Sturges* v. *Bridgman*. In this action a confectioner had for more than twenty years used a pestle and mortar in his back premises, which abutted onto the garden of a physician. The noise and vibration were not felt as a nuisance and were not complained of by the physician. In 1873 the physician erected a consulting-room at the end of his garden, after which time the noise and vibration became a nuisance to him. He accordingly brought an action for an injunction. It was held that the defendant had not acquired a right to an easement of making noise and vibration, and the injunction was granted.

In setting out the general principles when considering a nuisance claim, it was stated in *Sturges* v. *Bridgman* that what may be considered a nuisance in Belgravia might not be one in Bermondsey! Although this particular example is today rather unpalatable, the general rule is still applicable. The basic principle is that if you live in the middle of a city you cannot expect the same level of peace, cleanliness and air quality as you would if you lived in the country. However, in the country, the locals might be expected to put up with a cockerel crowing as the sun rises!

There have been a number of recent legal decisions in which the court has considered whether or not a nuisance affecting the use and enjoyment of a property is actionable, in the context of the nature of the locality in which the plaintiff lives. Essentially, these cases are restating the principle in *Sturges* v. *Bridgman*. In some cases, defendants have argued that the grant of planning permission has changed the nature of the locality and thereby extinguished the plaintiff's right to bring a nuisance action. Thus, the grant of planning permission to allow a factory to operate close to a residential area may change the locality of the area to the extent that a court might consider it was reasonable for a plaintiff to have to put up with a certain degree of disruption such as factory noise, smells and smuts. If this were the case, it would seriously restrict the potential for nuisance actions to protect individuals from the ill effects of such operations.

In the 1995 decision of *Wheeler and Another* v. *J. Saunders Ltd*, the Court of Appeal considered a case where planning permission had been granted to two pig-rearing houses. The plaintiffs brought a successful nuisance action as a result of foul smells from the unit. The question on appeal was whether the planning permission gave the defendants a response to the action. The Court of Appeal refused this defence and upheld the trial judge's award of damages and an injunction. The effect of that decision is that it confirms that a grant of planning permission does not authorise the commission of a nuisance. However, the question of whether a grant of a planning permission can change the character of an area to the extent that it will make it more difficult to bring a nuisance claim has not yet been fully resolved.

Thus, it is all the more important that local residents play an active role in the planning process to highlight what they consider the environmental problems might be before planning permission is granted, although this point is not directly relevant to the issue of what constitutes an actionable nuisance.

Continuation or repetition of incidents and time of occurrence
In the case of *Bamford* v. *Turnley* (1862) the court stated:

> What may be a nuisance in Grosvenor Square which would be none in Smithfield Market, that may be a nuisance at midday which would not be done at midnight, that may be a nuisance which is permanent which would be no nuisance if temporary or occasional.

Applying this principle, generally speaking there must be an element of continuity in the nuisance, and a single act will not normally amount to a nuisance. Temporary or occasional interference involving little actual or potential harm will rarely constitute a nuisance. The case of *Bolton* v. *Stone* (1949) involved a person who was hit by a cricket ball in her garden who tried to sue in nuisance. The Court of Appeal held that the claim failed because balls had been hit outside the ground only six times in the last 30 years.

Use of best practical means to remove the nuisance
The court will take into account whether the defendant has used the best means available to eliminate or reduce the inconvenience to the plaintiff. A finding that the defendant is not using the best means available, even if not conclusive, may well render the defendant liable. On the other hand, the fact that the defendant has used the best means available is by no means conclusive in the defendant's favour (save in the case of building and construction works, or where the defence of statutory authority is available to the defendant). In that case, the court would balance the degree of nuisance against the activities being undertaken and whether any steps could be taken to further minimise the nuisance.

In the case of *Read* v. *Lyons* (1947), the court stated that if a man commits a nuisance it is no answer to his neighbour that he took the utmost care not to commit it. A factory may be operated with reasonable care but the courts may still decide that it is a nuisance to its neighbours. The point here is that when the problem is considered from the plaintiff's viewpoint, the use of the land may be deemed unreasonable, even when having been balanced against all the other factors, including the nature and circumstances of the defendant's activity, the nature of the locality, best practicable means and social value and the level of interference to the plaintiff.

Social value of defendant's operation

The social value of the defendant's operation is a factor that the court will take into account. Fortunately for plaintiffs, it is not given much weight, otherwise an action against a factory employing local people and undertaking a process deemed useful to an industrialised society would always be a strong argument that the social value of the undertaking outweighs the nuisance to the individuals affected. In the case of *Adams* v. *Ursell* (1913), the defendant ran a fish and chip shop, which was causing a nuisance to the plaintiff because of the smells and noise emanating from it. The defendant claimed that he was performing a public service; however, the court held that there was no defence to the claim.

Foreseeability of damage

For the plaintiff to show that a nuisance has occurred, it must also be proved that the damage was foreseeable. In 99 per cent of cases this will not be a controversial point. In more complex cases – for example, the possibility of damage from methane gas caused by landfill sites – it may need to be proved with the assistance of expert evidence regarding the state of knowledge at a particular time.

In the 1994 case of *Cambridge Water Company* v. *Eastern Counties Leather plc* (see Chapters 12 and 13, pages 294 and 336), the defendants ran a leather-tanning plant which had used chemicals as part of the de-greasing process until about 1976. The plaintiffs had purchased a mill and constructed a water borehole in the area to extract drinking water. It found that the levels of organochlorine compounds had reached levels that made the water unfit to drink. The source of the pollution was traced back to the defendants' premises where spillages had occurred over many years. Chemicals had seeped into the underlying chalk aquifer. The plaintiffs had to set up another borehole at very considerable cost.

The defence to the action succeeded on the basis that the seepage and damage caused by the chemicals to the water had not been foreseeable at the time it occurred.

DEFENCES TO NUISANCE ACTIONS

Statutory authority

Once nuisance is established by a plaintiff, the defendant may try to raise the defence of statutory authority on the basis that the works form part of a statutory scheme. A 'special act' of Parliament enables a prospective defendant to raise the defence, whereas an act at the other end of the spectrum – for example, a grant of planning permission – does not. This defence causes particular problems when trying to establish an actionable nuisance against, for example, railway operators.

The reasoning behind the defence is that Parliament has authorised certain works that need to be done, usually for the common good, and Parliament's intention should not be thwarted. However, even if the defendants are able to raise a statutory authority defence, they must still show that the works were reasonably carried out and that all reasonable and proper steps were taken to ensure that no undue inconvenience was caused. The court will look at the methods and equipment used and have in mind the scale, value and duration of the works as a whole. It is up to the defendants to show that the works have been carried out with due diligence, and it will only be if they fail in this endeavour that the plaintiff will have a remedy.

It will be rare for the operators of an industrial process to show that they are entitled to rely on the defence of statutory authority. This will more often be the case in road and other transport schemes causing severe nuisance to those people affected. Most cases of nuisance caused by transport schemes will be dealt with by way of a statutory compensation scheme, most commonly under the Land Compensation Act 1973.

The House of Lords considered the defence of statutory authority in the case of *Allen* v. *Gulf Oil Refining Limited* (1981). In that case, it was held that the 'special act' of Parliament authorising the construction of the refinery conferred immunity on the company so that, as long as it could show that the nuisance being caused was an inevitable consequence of the works, there would be no entitlement to damages for nuisance. In these circumstances, the only occasion when a claim for nuisance can be made out is when the level of nuisance exceeds that which is inevitable. Therefore, if it is possible to show that all reasonably practicable steps were not being taken to minimise the nuisance, then a successful action might be brought.

It is clear, therefore, that once a defendant is able to show a defence of statutory authority, it becomes much more difficult for a plaintiff to bring a claim for nuisance. However, as this restricts a plaintiff's right to bring legal proceedings, the courts will normally be reluctant to infer a statutory authority defence except where it is clear that it exists.

Whether or not a defence of statutory authority can be relied on must be assessed on the individual facts of each case. In the recent Court of Appeal decision in the case of *Hunter and Others* v. *Canary Wharf Limited* (1995), it was clear that the courts are not willing to shut out plaintiffs from taking proceedings for nuisance based on an extension of the statutory authority defence principle.

Construction works

The courts recognise that it is almost inevitable that nuisance will be caused as a result of construction works, but because of their temporary duration, they are afforded a special protection from nuisance actions. In a construction nuisance claim where, for example, individuals are claiming that they have been affected by noise, dust and unreasonable working hours, they need to show that the nuisance has gone over and above that which was inevitable from the works had the defendants used the best practicable means to minimise the level of nuisance.

This makes it more difficult to bring a successful claim of nuisance in a construction works claim, but, where it is possible to show that inadequate methods of nuisance prevention have been used on the site, it is by no means impossible. Local authorities have power to control building sites and these are referred to in the section on statutory nuisance later in this chapter (page 358).

Previously existing nuisance

It is often mistakenly thought that it is a defence to argue that the nuisance was going on before the plaintiff arrived. However, this is not a defence. In the case of *Miller* v. *Jackson* (1977), the plaintiff moved to a new residential estate which bordered a long-standing cricket ground, from which balls were intermittently hit into the plaintiff's garden. The argument that this had happened before the plaintiff arrived failed to provide a defence to a nuisance action.

REMEDIES

One of the most useful aspects of a claim for nuisance is that it is possible to apply to the court for an injunction to prevent the nuisance recurring. It is more often than not the case that the reason for this type of action being brought by local people is to prevent further nuisances occurring in the future. However, it is not easy for an individual to obtain an injunction against a large industrial concern. The claim will have to be very strong if the court is to make an order that may effectively close down part of the operations. Understandably, a court would be reluctant to make such a harsh order in the absence of very strong evidence as to the degree of the nuisance and the steps being taken by the defendants to prevent it.

DAMAGES

It is possible to claim compensation for the past effects of nuisance. It is not entirely clear whether it is possible to bring a claim for damages for personal injury within the context of a nuisance claim. As the law in the area is undecided, most solicitors will include a claim for negligence in the alternative to a nuisance claim, where personal injuries are alleged as being caused by the nuisance.

Damages in cases where the use of enjoyment of property has been affected are usually modest. The most recent guidance came from the Court of Appeal in 1975 in the case of *Bone* v. *Seal*. In that action, the defendant owned a pig farm from which offensive smells emanated from time to time over a period of twelve years, giving rise to an actionable nuisance claim. At first instance, the judge calculated the damages at £500 per year, making a total of £6,000 (equivalent to about £31,400 today). The Court of Appeal reduced the amount to £1,000 for the whole period of twelve years' nuisance (at today's rate, worth about £5,200).

The Court of Appeal indicated that the level of damages in personal injury actions would provide a bench mark in assessing what the level of damages should be in a nuisance action. In the case of *Emms* v. *Collier* (1973) an award of £350 was made for one year's disruption resulting from building works. An equivalent sum at today's value is £2,400.

It is apparent from the levels of compensation that the courts will not award large sums where a plaintiff is suffering from a nuisance. In deciding the level of damages, the court will look carefully at the level of the nuisance, awarding greater sums where it is continuous rather than intermittent. The fact that the damages awards in nuisance cases are modest only serves to heighten the fact that the plaintiffs are usually motivated to bring actions for nuisance to prevent further occurrences of the nuisance, rather than for monetary gain.

ESCAPES OF DANGEROUS SUBSTANCES

A special principle of law has evolved from the law of nuisance which relates to the escape of dangerous substances. The principle is that a person who brings something onto his/her land that is likely to cause damage if it escapes is under a strict duty to prevent an escape. If that person fails, he/she is strictly liable to any person suffering loss or damage from the escape, whether or not the plaintiff can prove that the escape was caused as a result of negligence. This is known as the rule in *Rylands* v. *Fletcher* after a nineteenth-century court case.

The following are essential ingredients in a claim under the rule of *Rylands* v. *Fletcher* (1868).

- There is an accumulation on land of a 'substance' that is dangerous if it escapes. In previous cases, this has been held to include water, sewage, fire, gas, electricity, oil, acid smuts, fumes, explosives, animals and vibrations.

- The storage of the escaped substance constitutes a non-natural use of land. This has been open to interpretation by the courts, so that a munitions factory in wartime was then deemed as a natural user of land, whereas it was held in a recent House of Lord's case that the storage of quantities of chemicals on industrial premises constitutes a non-natural use of land, thus implying that the principle of *Rylands* v. *Fletcher* could apply to all industrial premises.

- The rule will only apply where there is an escape from land. The rule will not apply where the escape is caused by an act of God or the act of a stranger over whose acts the defendant has no control, or where the escape was due to the plaintiff.

In order to recover damages, the plaintiff must also show that the damage was foreseeable. It was previously unclear whether a plaintiff had to show that *some* damage was foreseeable or that the type of damage that actually occurred was foreseeable. The House of Lords decision on foreseeability, in *Cambridge Water Company* v. *Eastern County Leather plc* (page 352 above) applies equally to cases brought under the rule in *Rylands* v. *Fletcher*.

The rule in *Rylands* v. *Fletcher* is useful where there has been an emission or escape from industrial premises that has caused damage. The plaintiff can immediately put to one side the need to prove that the escape was caused by negligence, or that the defendants did not do enough to prevent the escape. This avoids

the need to embroil the case in complex and expensive technical
evidence as to what was going on in the plant that led to the
escape. Instead, the plaintiff can assume that the defendant will be
liable and needs only to prove a causal link between the damage
sustained and the escape, and that this damage was foreseeable.

Although the case of *Rylands* v. *Fletcher* appears to be a really
useful additional claim, the reality has in fact been that very few
cases have ever succeeded where it has been pleaded. It is more of
a theoretical head of claim than a real one.

STATUTORY NUISANCE CLAIMS

Part III of The Environmental Protection Act 1990 (EPA) is entitled
'Statutory Nuisances and Clean Air' and deals with statutory
nuisances, clean air and controls over offensive trades. Sections
79–82 replace the previous provisions of the Public Health Act 1936
with a more streamlined system of procedures to allow local authorities to deal with nuisance. The local authority will usually be the first
port of call when an individual considers that a nuisance is occurring, and the provisions under the EPA set out the types of nuisance
that the local authority has power to deal with, and what action they
are required to or have a discretion to take.

For an individual with an actual or potential nuisance problem,
it is useful to know what the local authority can do under the
provisions of the EPA and legislation that empowers it to act in
given circumstances. Local authorities often have a host of reasons
for failing to act in order to prevent nuisance. Knowing what
powers are available to local authorities can be a useful device in
trying to persuade officers or members that they are failing to
abate a nuisance within their area when they have the statutory
power to take that step.

Duty to investigate and follow up complaints

Under section 79(1), every local authority is placed under a duty to
ensure that its area is inspected from time to time to detect any
statutory nuisance. It is also required to take such steps that are

reasonably practicable to investigate a complaint of statutory nuisance made by a person living in the area. Total inaction would not satisfy the requirement, which is also unlikely in the case of a once-only inspection. Where a local authority fails to satisfy the requirements, the Secretary of State for the Environment, Transport and the Regions has powers to take action.

Definition of statutory nuisance

A statutory nuisance is defined by section 79(1) of the EPA as meaning:

(a) any premises in such a state as to be prejudicial to health or a nuisance;

(b) smoke emitted from premises so as to be prejudicial to health or a nuisance;

(c) fumes or gases emitted from premises so as to be prejudicial to health or a nuisance;

(d) any dust, steam, smell or other noxious substances arising from an industrial trade or business premises and being prejudicial to health or a nuisance;

(e) any accumulation or deposit which is prejudicial to health or a nuisance;

(f) any animal kept in such a place or manner as to be prejudicial to health or a nuisance;

(g) noise emitted from premises so as to be prejudicial to health or a nuisance;

(h) any other matter declared by any enactment to be a statutory nuisance.

The sections state that in respect of each class of statutory nuisance, the nuisance must be 'prejudicial to health or a nuisance'.

Section 79(7) defines 'prejudicial to health' as meaning 'injurious or likely to cause injury to health'. This is the same definition as used by the 1936 Public Health Act, and it makes it clear that likelihood of injury to health is included as well as actual injury. However, nuisance is a separate test and it is not necessary to show that the activity is prejudicial to health in order for it to be a statutory nuisance. It has also been held that a nuisance in terms of what could be defined as a statutory nuisance has to be given a common-law interpretation – that is, that the cases be deemed to arise out of civil law, private or public nuisance as appropriate.

The definition 'prejudicial to health' could also be widely interpreted by the courts, particularly if it is seen in the light of the World Health Organisation's definition of health. This states that health is 'a state of complete physical, mental and social wellbeing and not merely the absence of disease or infirmity'. So far, however, the English courts' interpretation of 'health' has been narrow, although, with increased health and environmental awareness, there is scope for broadening this.

Here are examples of cases where the issue of health has been addressed:

- In *Bennett* v. *Preston Borough Council* (unreported), the judge held that unsafe electrical wiring could contribute to premises being prejudicial to health, if not from the threat of electric shocks and fires, at least from the smoke that would result from a fire.

- In *Day* v. *Sheffield City Council* (unreported), the court held that a caravan site was prejudicial to health by reason of rodent infestation, open skips for refuse and an uncovered drainage gully below a water tap, lack of site lighting and uneven surfaces making it unsafe to cross the site at night, and the presence of barbed wire on site.

- In *Greater London Council* v. *Tower Hamlets London Borough Council* (1983), the court held that condensation and its associated mould growth could render premises prejudicial to health.

- In *Southwark London Borough Council* v. *Ince* (1989), premises were regarded as prejudicial to health by reason of noise from road and rail traffic penetrating the dwelling due to inadequate sound insulation.

'Best practicable means' defence

The defence of 'best practicable means' is set out in section 79(9) of the EPA and it applies to most, but not all, of the categories of statutory nuisance. This is where the statutory system differs from a claim for nuisance under the civil law. The definition of 'best practicable means' is set out in section 79(9), where it is interpreted by reference to the following:

(a) 'practicable' means reasonably practicable having regard among other things to local conditions and circumstances, to the current state of technical knowledge and to the financial implications;

(b) the means to be employed include the design, installation, maintenance and manner and periods of the operation of plant and machinery, and the design, construction and maintenance of buildings and structure.

The test only applies insofar as it is compatible with safety issues and other duties imposed by law. In circumstances where a code of practice has been issued under section 71 of the Control of Pollution Act (COPA) 1974, regard should be had to any guidance given. The court will take into account the question of cost in deciding whether a method of reducing the nuisance is practicable. The mere fact that this would increase expenditure or reduce profitability would not be sufficient to establish the defence. If a company intends to rely on the defence of 'best practicable means', it must bear the onus of proving compliance with this concept.

Procedure for bringing a statutory nuisance claim

If a local authority considers that a statutory nuisance is occurring, it can serve an abatement notice (that is, a notice asking that those causing the nuisance desist). The abatement notice can impose all or any of the following requirements:

- the abatement of the nuisance by prohibiting or restricting its occurrence;

- the requirement for work to be carried out as may be necessary for this purpose;

- specification of a time within which the requirements of the notice should be complied with.

The notice has to be served on the person who is responsible for the nuisance, stating that the person served has the right of appeal and specifying the precise nature of the nuisance complained of and what is required to prevent a recurrence. If the people causing the nuisance either ignore the abatement notice or contravene it without reasonable excuse, they will be guilty of an offence. This will expose them to liability for a fine or conviction in the magistrates' court. If the nuisance is committed on an industrial trade or business premises, the maximum fine has been increased to £20,000.

Who can claim?

Although the EPA envisages that the local authority will be the prosecutor, section 82(1) allows any person 'aggrieved by the existence of a statutory nuisance' to make a complaint to the magistrates' court. The party does not have to be an occupier of premises, but merely aggrieved by the existence of the nuisance. The definition has been widely interpreted in the past and is likely only to exclude 'busybodies' who are not themselves affected by the act/premises complained of.

In a recent case, *Herbert* v. *Lambeth London Borough Council* (1991), the court held that a finding of statutory nuisance amounted to a conviction, thus allowing the magistrates' court to make a compensation order. Also section 82(12) provides that costs must be awarded in favour of a complainant where the nuisance existed at the date of the complaint, whether or not it has been subsequently abated.

Claims against local authorities

Local authorities need to bear in mind that the provisions of the EPA can also be used against them. This principle was well established in relation to section 99 of the Public Health Act 1936, particularly for cases requiring the local authority to carry out repairs or defects to housing.

OTHER PROVISIONS DEALING WITH NOISE NUISANCE

There are other specific provisions governing noise that are set out in Part 3 of the COPA 1974. For example, sections 60 and 61 are aimed at allowing local authorities to control noise and vibration from construction sites. A local authority may serve a notice under section 60 imposing requirements concerning how construction works are to be carried out. The notice may specify:

- the type of plant and machinery to be used or not to be used;

- the hours to be worked;

- noise levels for the site.

The notice will usually require that best practicable means should be used on site to minimise nuisance from the noise, and in serving the notice the local authority will usually have a guide to the relevant provisions of a code of practice – in particular, the code of practice for open sites.

Section 61 of the COPA allows the developer to agree on noise requirements and in advance of starting work. Thus the contractor or developer will agree noise requirements with the local authority in advance of the work commencing. The local authority has the power to attach conditions to the consent given under section 61.

CONCLUSION

In conclusion, where a person can show that they are suffering a nuisance, a civil claim in private or public nuisance will allow them to claim an injunction to stop the nuisance and damages for loss of enjoyment of their home. Legal aid is available to people who satisfy the merits and financial eligibility tests (see Chapter 3).

It is still unclear whether the remedy of private nuisance is available to those without a proprietary interest. If the House of Lords decides in favour of the plaintiffs in the Docklands case, the courts will no longer make the distinction between owners/tenants and those who simply *occupy* a place as their home. However, if the House of Lords accepts the defendants' arguments then the potential scope that actions for nuisance afford ordinary people will be severely limited. This would mean that the law of nuisance would be a property-based right, rather than one that could be used to protect the quality of one's environment. If that is the case, then individuals may still be able to claim under public nuisance but only if they can show that they are more particularly affected than others who are experiencing the same problem.

Generally, the level of damages awarded by the courts for nuisance claims has been relatively modest. Damages are assessed on the degree and level of the nuisance suffered. The modest level of damage awards makes it difficult to obtain legal aid for individuals, and it is often necessary for a group of people affected to bring a claim in order to make it economically viable. However, in most cases, injunctions are sought as the appropriate remedy, since those seeking redress are usually more interested in stopping the occurrence of the nuisance than in monetary compensation.

Alternatively, statutory remedies are available under the Environmental Protection Act 1990 (EPA) where it can be shown that the nuisance comes within the range of offences defined in the EPA. In most cases, an individual will need to enlist the cooperation of his/her local authority, as the EPA's provisions are generally geared towards situations where the local authority will bring the claim on the individual's behalf in the magistrates' court. If the claim is successful, the local authority will enforce the magistrates' order by serving an abatement notice on the defendant in respect of the nuisance to lessen its effects or to stop it altogether, with penalties attached if the defendant fails to comply with the terms of the abatement notice.

Local authorities are required to take such steps as are reasonably practicable to investigate a complaint of statutory nuisance made by an individual living in the area. Further redress may be sought if the local authority fails in its duty to do so (see Chapter 9 on judicial review).

The EPA also gives scope for individuals in their capacity as 'persons aggrieved' to represent themselves by making complaints directly to the magistrates' court. Unlike civil claims in private nuisance, therefore, the issue of claimants needing a legal interest in land before they can claim does not arise. In practice, individuals are often less keen to represent themselves before magistrates, because legal aid is not available.

15 Water Cases

This chapter does not attempt to provide a comprehensive overview of all aspects of water law. Instead, it focuses upon the particular areas related to water pollution that have most direct impact upon members of the public and that are most accessible to legal action being taken by individuals and by environmental action groups.

PROSECUTIONS

Criminal prosecutions for water pollution offences have been, to date, almost exclusively within the remit of the Environment Agency and its predecessor the National Rivers Authority (NRA). Although, theoretically, anyone can bring a prosecution for the offence of polluting controlled waters (contrary to section 85 of the Water Resources Act 1991), in practice this has rarely happened.

Some particular interest groups, such as the Anglers' Association, have successfully brought prosecutions on behalf of their members where pollution of rivers has resulted in wide-scale loss of fish stocks. The courts have powers to order persons convicted of polluting controlled waters to restock fish supplies.

Common-law actions are also available to riparian owners – that is, owners of land through which rivers run. At common law, landowners have a right of action against anyone polluting the water running through their land.

Successful actions have been brought in nuisance, both private and public, trespass, negligence and by way of the rule developed in the case of *Rylands* v. *Fletcher* (1868), a form of extension of nuisance whereby a person can be liable for damage when there is an escape of a dangerous substance from his/her property. This liability can be strict if it arises from a non-natural use of land (see Chapter 14, pages 356–8).

Shellfishers are another group with particular and specific concerns about coastal and estuarine water pollution. In recent years members of this specialist group have become involved in a number of actions against water companies for sewage contamination of their harvest. Actions have been brought in common law, private nuisance and trespass, and also by way of section 7 of the 1967 Shell Fisheries Act.

PERSONAL INJURY

Over the last few years a number of actions have been brought by individuals claiming damages for personal injuries suffered as a result of exposure to bacteria and viruses present in the sea water. These actions have predominantly been taken by surfers, windsailers and swimmers. As yet, none of these cases has proceeded to trial.

A particular difficulty in pursuing these cases, in negligence, is that of establishing causation. By definition, most bacteria and viruses present in sewage are also present in the environment. In common-law negligence actions, the burden of proof falls upon the individual bringing the action, who must prove that the sewage was the source of the infection as opposed to anything else. As will be appreciated, this can be a very high hurdle to overcome.

The case studies set out below concentrate on the quality of drinking water, which potentially affects every person in the country, and the state of beaches and bathing water in the UK which are used by roughly 20 million people every year.

As will be seen, contamination of drinking water and pollution of beaches and bathing water are relatively commonplace in the UK and have been the source of successful legal actions brought by both individuals and environmental groups.

DRINKING WATER

The Camelford case

In July 1988 a delivery driver made a routine delivery of 20 tonnes of aluminium sulphate to a water-treatment works at Lowermoor in north Cornwall. The water-treatment works was owned by South West Water Services Limited and was responsible for providing drinking water to an area of north Cornwall centred on the town of Camelford.

Through a combination of circumstances, the aluminium sulphate was delivered into the wrong inlet and, as a result of this, it mixed with water in the distribution system that was for domestic purposes, including drinking. The effect of mixing aluminium sulphate in water is to produce a form of dilute sulphuric acid, together with residues of aluminium.

Within a relatively short space of time large numbers of people in the Camelford area began to exhibit a wide range of symptoms, both acute and chronic. The effects included aching limbs, fatigue and general malaise, but more seriously there was also evidence of minor to moderate brain damage, taking the form of short-term memory loss, difficulties with concentration and some intellectual impairment.

In January 1991, South West Water Services Limited were convicted at Exeter Crown Court of the criminal offence of causing a public nuisance by endangering the health and comforts of members of the public through their failure to supply wholesome water.

That same year, approximately 200 individuals who had suffered ill effects from the contaminated water commenced civil proceedings against, primarily, South West Water Services Limited who were responsible for supplying them with drinking water. The cases were eventually settled, out of court, just before trial.

The Camelford incident provides a very good illustration of how a multiplicity of legal causes of action can stem from a single, relatively straightforward event.

Actions at common law were brought in negligence, alleging a wide range of failures on the part of the water company; for liability arising under the rule in *Rylands* v. *Fletcher* (on the basis that the water company had allowed the 'escape of a dangerous substance' from its custody – see Chapter 14, pages 356–8); and by way of breach of contract, on the basis that it was possible to imply a term that water supplied by the company should be wholesome and free from defect. In addition, various breaches of statutory duty derived from both national and European law were also pleaded.

It was claimed by the plaintiffs that the water company had supplied a 'defective product' as set out in the Consumer Protection Act 1987. This issue was stressed in the case of *Gibbons* v. *South West Water Ltd* (1993), where the court held that domestic water was a product for the purposes of the Act and, therefore, fell within its scope.

It was also claimed that the water supplied was in breach of various levels as set out in the European Union Water Quality Directive (80/778/EEC). This Directive sets quality standards for a wide range of parameters to which human consumption should conform. The Camelford residents claimed that the water failed to satisfy standards in respect of aluminium, lead, copper, zinc, iron and sulphate.

In addition, it was claimed that the domestic statute in force at the time, the Water Act 1989 (now replaced by the Water Industry Act 1991 and the Water Resources Act 1991) had also been breached in that the water company had failed to take all reasonable care to provide a supply of wholesome water sufficient for domestic purposes.

Other drinking water cases

At the present time there are a number of similar cases in the offing. Each of these actions involves tens, if not hundreds, of people and it is most likely that, provided people can prove the cause of their illness, the water companies responsible for supply will settle out of court in order to avoid the ever-increasing costs of litigation and the attendant bad publicity.

Potential and actual sources of contamination of drinking water in the UK are rife. Aside from the Camelford incident, many other people have had their drinking water supplies affected by environmental contaminants.

In East Anglia, pesticide residues have allegedly found their way into water supplies, through the contamination of ground water. It is being argued that there have been numerous outbreaks of poisoning from the bacterium cryptosporidium as a result of sewage contamination of drinking water. Another growing problem is the contamination of ground water from heavy metals being leached from landfill sites. An increasing concern for communities in former mining areas is the contamination of ground water from abandoned mines.

In addition, recent concern has focused on the contamination of drinking water supplies by a wide range of industrial solvents, such as trichloroethylene and tetrachloroethylene, both of which have found their way into drinking water supplies. At the present time there is also growing concern about the likely presence of endocrine disrupters (substances that affect the working of the hormone production system in the human body) in drinking water, which, it is thought, may have a particularly adverse effect on human fertility. This is considered in part to be related to the presence of residues of oestrogen passing into the drinking water system as a result of women taking the contraceptive pill. In the first instance individuals would again be advised to contemplate action against the drinking water supplier.

The basis for standards regulating the quality of drinking water is the EU Drinking Water Directive referred to above, which sets out quality standards for 'water intended for human consumption'. It provides parameters for a wide range of possible contaminants, such as heavy metals, bacteria and chemicals. In most instances the Directive requires that certain maximum concentrations should not be exceeded. The Directive also sets out a number of monitoring and sampling requirements.

The Directive is largely implemented in the UK by the Water Supply (Water Quality) Regulations 1989 and is further implemented by the Water Industry Act 1991. Section 68 of the Water Industry Act imposes a duty upon water suppliers – that is, the privatised water companies – to supply domestic consumers with 'wholesome' water, as defined in the 1989 Regulations.

It is a criminal offence not to supply wholesome water or to supply water that is unfit for human consumption. However, the legislation provides that proceedings may only be taken by either the Secretary of State for the Environment (now Environment, Transport and the Regions), on the recommendation of the Drinking Water Inspectorate, or by the Director of Public Prosecutions. By way of section 210 of the Water Industry Act 1991, the directors of the supply companies can be held personally liable.

It is worth noting, however, that it was not until April 1995 that a water company was first convicted of supplying water 'unfit for human consumption'. Severn Trent Water was fined £45,000 at Hereford Crown Court following the contamination with solvents of water supplies to a large number of homes in the Worcester area. It is thought that up to 60,000 households were affected by this.

Following the Camelford case it also appears that the supply of water not up to standard may also be actionable through the common law of public nuisance. Product liability claims are also possible, but, although the claim was made against South West Water in the Camelford case under the Consumer Protection Act 1987, it is not possible to plead an action under the Supply of Goods Act 1979 because it appears that water does not satisfy the definition of 'goods' in the Act, nor is it strictly supplied under a contractual arrangement but rather under the duty to supply as set out in section 68 of the Water Industry Act 1991.

The UK government has fallen foul of the Drinking Water Directive on numerous occasions, actions being largely prompted by complaints made to the European Commission by, in particular, Friends of the Earth.

In 1989 the Commission began proceedings against the UK government for excess nitrate levels in drinking water supply in the Norwich area, such action being provoked by the fact that there was evidence to suggest that the Directive had been breached more than 100 times in different areas between 1985 and 1989.

In 1992 the European Court of Justice ruled that the UK government was in breach of the Drinking Water Directive and that, furthermore, the duty to comply with the Directive was an absolute one, and not conditional upon simply taking practical steps towards compliance.

The Drinking Water Directive also sets out maximum concentrations for lead in drinking water. Once again, various surveys showed significant breaches of maximum concentrations in a very high proportion of UK households. It is likely that the number of cases arising from failure to provide drinking water that complies with the Directive will, if anything, increase in the future.

BEACH POLLUTION

The Croyde Bay case

Croyde Bay is situated on the northwest Devon Coast close to the holiday resorts of Woolacombe and Saunton. It is enclosed by two headlands and has a broad, sandy foreshore backed by sand dunes. The bay is designated as an area of outstanding natural beauty and is also a site of special scientific interest. In addition, the waters of Croyde Bay are designated as bathing waters for the purposes of the EU Directive Concerning the Quality of Bathing Water (76/160/EEC). (For more on the Bathing Water Directive see Chapter 6, pages 101–5.)

In April 1988 a local businessman, Michael Saltmarsh, purchased an area of 180 acres of land at Croyde. The land included a caravan park, a campsite and a number of buildings that were rented out as holiday accommodation. Unusually, the site also consisted of the entire foreshore of the beach, down to low water mark. As a general rule, land lying seaward of high water is the property of the Crown, but at Croyde, the freehold of the foreshore had been granted to a local family in the thirteenth century, and had remained in the hands of the same family until Mr Saltmarsh's purchase.

The existing site had become rather run down over the years: Mr Saltmarsh planned to develop the site as a holiday resort and to convert a number of the existing buildings into time-share apartments.

Very shortly after the purchase of the site, Mr Saltmarsh realised that the beach had a particular pollution problem – regular deposits of sewage-related debris. Each new tide brought in quantities of sewage-related plastics, such as panty-liner backing strips and condom rings, together with the more usual marine flotsam and jetsam.

The source of the problem was immediately apparent. Approximately one kilometre to the north of the beach along a rocky headland was an unscreened raw sewage outfall which served the parish. Although the winter population was only in the region of 2,000, in the summer this would swell to over 10,000, leading to enormous amounts of sewage being pumped out through this outfall.

To add insult to injury there was also a storm overflow, which discharged from a sea wall directly onto the foreshore of the beach. The path of sewage could be traced across the beach by a trail of heavy algal growth due to nutrient enrichment.

Although Mr Saltmarsh cleaned his beach daily, his business began to suffer as a consequence of growing public awareness of the health risks of sewage contamination. As the owner of one of the best surfing beaches in North Devon, Mr Saltmarsh was put under increasing pressure by environmental groups such as Surfers Against Sewage, who were concerned about the risks to health at Croyde Bay.

Apart from the damage to his business, Mr Saltmarsh became increasingly concerned about his personal liability in the event of bathers becoming ill. He was faced with the dilemma of either warning beach users about the possible health risks, thus further damaging his business, or remaining silent and exposing himself to possible liability.

Mr Saltmarsh approached the National Rivers Authority (NRA) as the statutory body then responsible for the issue of consents to discharge and for controlling pollution of coastal waters. The position of the NRA was that a consent to discharge had been issued with which, on the face of it, the water company responsible, South West Water Services Limited, appeared to be complying and there was therefore no action that the NRA was prepared to take. The local district council took the view that Mr Saltmarsh's problems were beyond their remit and they had no powers to take any action.

The position of the water company operating the outfalls was that, as confirmed by the NRA, they possessed legal consents to discharge and there was no legal obligation for them to upgrade the level of treatment of sewage discharged into Croyde Bay.

Faced with the prospect of no assistance from the statutory authorities, no improvement from the water company and continuing business losses, Mr Saltmarsh sought legal advice as to whether there were any remedies available that would both compensate him for his financial losses and stop the pollution of his beach.

Mr Saltmarsh's solicitors advised that could take action against the water company responsible for the pollution using common-law negligence, private and public nuisance and trespass.

The negligence alleged was that the water company had breached their duty of care to him by failing to take reasonable steps to ensure that the discharge of sewage and sewage-related debris from their outfalls would not cause damage to Mr Saltmarsh's land. As the freehold owner of the foreshore, Mr Saltmarsh was also able to proceed in private nuisance and in public nuisance (see Chapter 14). In addition, his ownership of the foreshore allowed him to proceed in trespass.

When court proceedings were issued, the schedule of financial loss – the document setting out details of his claim – included claims for the costs of clean-up (costs of both labour and plant), loss of profits from the business (both time-share and resort) and depreciation in the value of Mr Saltmarsh's property. The total claim amounted to about £2.5 million. In addition, Mr Saltmarsh claimed 'injunctive relief' – that is, an interlocutory injunction or court order – restraining the water company from performing any act that would result in the deposit of sewage-related debris on his property. (This injunctive relief was not in fact granted as the matter settled before the hearing, see page 375 below.)

The announcement of the commencement of proceedings by the issue of the writ was timed to coincide with the August bank holiday 1993, thus ensuring the case received maximum publicity. This was successful to the extent that the case was reported on national television and radio and the national press, including the front page of *The Times* newspaper.

As the case was primarily a nuisance action, Mr Saltmarsh suffering damage to his property and loss of enjoyment of it due to unlawful interference, it was listed to be heard in court by the Official Referee in the Queen's Bench Division of the High Court.

This judicial officer hears trials of cases where prolonged examination of documents may be necessary. The Queen's Bench Division is one of the three divisions of the High Court and it deals mostly with common-law actions. This ensured that the matter would progress more speedily than were the hearing to go to a judge, thus limiting costs and increasing pressure upon the defendants.

An order for directions as to how the case would proceed was made by the Official Referee in December 1993, fixing a trial date for June 1994. Experts as to liability or fault were to be limited for each side to three, namely, a civil engineer, a marine hydrologist and a microbiologist, while quantum experts considering the value of the case were limited to two, an accountant and a tourism economy expert.

The civil engineer was essential to provide evidence on sewage-treatment technology, the marine hydrologist was to provide evidence of the effect of tides, winds and currents upon sewage and sewage debris discharged from the outfall, and the microbiologist was to provide evidence on risks to health from sewage bacteria and viruses. A tourism economy expert was essential to provide evidence of financial loss, undertaking surveys of visitors to Croyde and neighbouring resorts, thus being able to provide good comparisons with neighbouring resorts that did not have problems with sewage pollution.

As the trial was not due to commence until June 1994, and the problems of sewage pollution continued, an application for an 'interlocutory injunction' – that is, an interim order preventing continuing pollution – was made in early April 1994. This was a means of both remedying the immediate problem pending trial, thus allaying concerns about the forthcoming holiday season, and further increasing pressure upon the defendants.

In the event, it was not necessary for the court to grant an injunction, as the defendants settled out of court prior to the hearing of the summons for the injunction, much to the satisfaction of Mr Saltmarsh. The terms of the settlement are the subject of a confidentiality agreement, although the fact that there is a settlement is not.

Since the settlement, South West Water has undertaken a number of improvements to the sewage system at Croyde, and the problem of sewage pollution of the beach appears to have improved significantly.

Use of judicial review – R. v. Carrick District Council ex parte Shelley

Porthtowan is a small village on the north Cornish coast, midway between Newquay and St Ives. The beach at Porthtowan is very popular among both locals and visitors as the beach is sandy, accessible to the public and particularly good for surfing. The area of the beach above high water mark is owned by the local district council, Carrick, while the foreshore is owned by the Duchy of Cornwall and leased to the National Trust.

For many years complaints were made to the local council, to the regional water company, South West Water Services Limited, and to the NRA about the problem of sewage-related debris being regularly deposited on the beach at Porthtowan.

Along the coast from Porthtowan are two major sewage outfalls. One serves the town of Redruth and discharges at Portreath. This was screened in 1991. The other, discharging at North Cliffs, serves the town of Camborne and was not screened. Both outfalls are operated by South West Water Services Limited and both require consents to discharge from the NRA (since April 1996 the Environment Agency). To discharge without a valid consent and to discharge in breach of a condition of a consent are offences under section 85 of the Water Resources Act 1991.

A consent was issued for North Cliffs by the NRA on 7 November 1992. One of the conditions of the consent was that the discharge should be screened by May 1994. An appeal against this condition was lodged by South West Water Services Limited to the Secretary of State for the Environment in March 1993.

By the time this case came to trial in April 1996 the Department of the Environment had still not determined the appeal. During the course of the hearing it emerged that an appeal against a consent condition does not have the effect of suspending the condition appealed against. It was therefore accepted in court that, strictly speaking, all discharges from the North Cliffs outfall since May 1994 had been unlawful as being in breach of the screening condition. The NRA, however, had adopted a policy not to enforce such conditions pending the outcome of any appeal to the Secretary of State.

In compliance with the litter provisions of the Environmental Protection Act 1990, the district council clean the beach at Porthtowan every day between May and September. The council keep records of the quantities of sewage-related debris collected.

During the summer of 1994, an average of 1.5 kilograms per day of sewage-related debris was collected, this figure falling to an average of about one kilogram per day for 1995. The sewage-related debris consisted largely of sanitary towels and condoms.

Over a period of time the district council received numerous complaints from local people, visitors, local businesses and the locally based Surfers Against Sewage. In 1993 the council was called upon by these groups to take action under section 80 of the Environmental Protection Act 1990 (EPA 1990) to serve an abatement notice against South West Water Services Limited to stop the discharge of debris.

Section 79 of the EPA 1990 imposes a duty on local authorities to investigate their area for the existence of statutory nuisances. It was suggested to the council that section 79(1)(e) of the EPA 1990, which refers to 'any accumulation or deposit which is prejudicial to health or a nuisance', was applicable to the sewage-related debris found at Porthtowan beach.

Under section 80 of the EPA 1990, where a local authority is satisfied that a statutory nuisance exists, or is likely to occur or recur, it 'shall' serve an abatement notice. The wording of the statute does not give a local authority a discretion, but imposes a mandatory duty to serve such a notice.

The culmination of the pressure exerted upon the district council was a meeting of the environment committee of the district council held in June 1995 to consider the possibility of action under section 80 of the EPA 1990.

The committee resolved that it did not consider it appropriate to take action under the EPA 1990 but preferred to continue monitoring the situation and to put further pressure on the Department of the Environment to reject the appeal of South

West Water against the condition of the discharge consent
requiring screening of the outfall.

This resolution of the council was clearly a decision that could
be challenged by way of judicial review, and two Porthtowan
residents, Rachel Shelley and Sara Delaney, who had young
children who used the beach practically every day, decided to
challenge the resolution.

Leave of the court to take a judicial review action was granted
in November 1995 and a full hearing was held in March 1996.
Together with Carrick District Council, the NRA and South West
Water Services Limited were also represented in court.

The crux of the applicants' case was relatively straightforward.
They claimed that by concluding that it was not appropriate to
serve an abatement notice, the environment committee had
misguided themselves as to whether they had a discretion to serve
a notice or not. The court made clear that the question of whether
to serve a notice or not was one of fact. Was there an accumulation
or deposit that either constituted a nuisance or was prejudicial to
health? If so, the decision-making body must then ask itself who
was responsible for creating the nuisance and serve a notice upon
them – there was no discretion.

Further, the council appeared to have thought that they could
defer any decision pending the outcome of the NRA appeal. The
court held that the statutory duty of a local authority is not
affected by the action of the NRA under the Water Resources Act
1991 but that it has a separate and independent duty.

The court also clarified a number of other issues in respect of
environmental statutory nuisances.

The decision to serve or not to serve a notice under section 80 is
based on whether, on the balance of probabilities, the council is
satisfied that a nuisance exists. It does not have to be satisfied
beyond reasonable doubt, the criminal standard of proof.

Again, when it comes to determining upon whom the notices
must be served, this too should be approached on common-sense
lines and the council needs to be satisfied on the balance of
probabilities who is responsible for the nuisance.

There was some discussion in court about the definition of
what constitutes a statutory nuisance and the judge held that

statutory nuisance fell within the field of public nuisance. The common-law definition of public nuisance should be adopted – that is, something that 'materially affects the material comfort and the quality of life of a class of Her Majesty's subjects'.

In addition, there was no requirement that a statutory nuisance needed to prejudice public health in any way and the judge accepted that as statutory nuisance is set out in the EPA 1990 it should therefore have a wider ambit, in the context of an environmental Act, than definitions derived from the Public Health Act 1936.

The judge had no difficulty in accepting that, in principle, significant deposits on a public beach of sewage-related debris were capable of constituting a statutory nuisance, even without specific evidence of injury to health.

In the event, the court held that the council had failed to discharge their duty as set out in Section 80 of the EPA 1990 and the local authority accepted that it was under an obligation to reconsider its previous decision.

Since the trial, South West Water has announced that, without the appeal being determined by the Department of the Environment, it will fit screens to the North Cliffs outfall. Notwithstanding this, the council served an abatement notice on the company in July 1996.

CONCLUSION

The chapter has demonstrated the variety of legal mechanisms available to remedy an existing state of affairs. As was seen by the Croyde Bay example, these mechanisms may be necessary despite the operation of a regime designed to prevent damage to the environment. Although in the Croyde Bay case there was an ongoing dispute as to the enforcement of a condition attached to a licence, nevertheless there was a legal consent to discharge in place.

16 Evidence: Building Your Case

This chapter is a guide to the significance of evidence for those involved in environmental litigation. It also offers some guidance for those charged with a criminal offence after involvement in a public protest.

The chapter assumes that a lawyer will be running the case, but aims to explain what a lay person can do to help, what steps are taken to obtain evidence at what stage and why. For those who are running cases for themselves, the chapter should act as a preliminary reference, but they should also refer to books such as *Pollution and Personal Injury Toxic Torts II* by Charles Pugh and Martyn Day (1995).

What is evidence?

Evidence is the means by which a party to a court case proves his/her version of the story. There are three main types of evidence: oral, documentary and real.

- Oral evidence is given by witnesses in the witness box (or sometimes by video link in the courtroom).

- Documentary evidence is documents that include information to be used by one side or other to prove their case. An enormous variety of documents are used. Most cases will have witness statements and experts' reports which will form the basis of oral evidence given in the witness box. However, depending on the nature of the case, documentary evidence can include letters of complaint, letters informing people of problems, medical records, accident reports at work or by the police, planning permissions, charts or graphs, contracts of employment, and so on – the list is endless.

• Real evidence is a specific object, such as photographs, a video, a diary, a knife, a piece of machinery, etc.

The phrase 'burden of proof' will be frequently heard by anyone who has contact with lawyers and the courts. The people upon whom the burden of proof falls have the responsibility of proving their side of the story, and if they do not have enough evidence to do this their case will fail. In criminal cases it is the prosecution or the Crown Prosecution Service (CPS) that bears the burden of proof, and in civil cases it is the plaintiff who has to show how the defendant has breached a particular law, thereby causing damage.

The amount of evidence needed to prove the case varies, depending on the type of case involved. In criminal proceedings, the prosecution has to convince the court 'beyond reasonable doubt' that the person committed the offence with which they are charged. In civil proceedings, the plaintiff has to convince the court 'on the balance of probabilities' that the defendants failed, for example, to provide their employee with adequate health and safety protection. Put another way, the 'balance of probabilities' means that it is 'more likely than not' that the defendant should have acted differently and can be held responsible for the consequences. As a rough guide, criminal proceedings require an 85 to 90 per cent level of proof and civil proceedings require more than a 50 per cent level of proof.

What makes good evidence?

First and foremost, all evidence must be relevant to the actual charges or allegations made by either party and the specific law under which the case is being brought. For example, a family living next to a chemical factory may overestimate the relevance of notes in their diaries listing numerous incidents of unpleasant smells in their home. In fact, to have a case in nuisance the family has to show that the smells they endure are out of the ordinary for the industrial nature of the neighbourhood; evidence of frequent smells may show nothing of the kind and may even confirm that the smells are just part of what can only be expected.

Once accepted as relevant, different pieces of evidence vary in their significance. The court has to decide what weight to attach to a particular witness's oral evidence, to a set of photographs or whatever. The best evidence has either one or more of the following characteristics: independence, precision or contemporaneity. Much evidence will not have all three. The fundamental question before the court is whether the total evidence provided by the party with the burden of proof is good enough to prove his/her case to the level required.

Independence

Independent evidence means evidence provided by a person who does not have a personal interest in the outcome of the litigation. For example, in a case where the plaintiff alleges that a building site made unreasonable noise at 6 a.m. on Sunday mornings, the local vicar walking past the site on his way to early morning service could provide good evidence that noisy building works were indeed carried out at that time. He is unlikely to have anything to gain by giving the evidence and is likely to give an impartial account.

Precision

Precise evidence is more convincing because it is less likely to be mistaken and can clearly detail a breach in the law. Using the same example, the vicar's evidence is more convincing if he says, 'I walked past the building site on 27 April 1996 at 6.10 a.m. with my dog. There were two men using a crane, one driving and the other shouting directions. I remember the time because I was late and kept looking at the church clock ...', rather than, 'One Sunday in 1996 when I was walking to the church for the early morning service I remember seeing some men working on the building site.' In the last version of the story, the vicar could be walking to church at 8.30 a.m. and the men working could have been in the office creating no noise at all.

Contemporaneity

Contemporaneous evidence is evidence that was created at the time of the alleged event. For example, the vicar may have written down the details of the incident in his diary on the evening of 27 April 1996. Producing the diary in court helps to demonstrate that his account of the events, which are likely to have taken place several months or years earlier, is not entirely reliant on his memory.

Admissibility

There are numerous legal rules about 'admissibility' of evidence. An inadmissible piece of evidence is not allowed to be put before the court. In criminal proceedings, for example, a confession could be inadmissible if it was obtained by threatening violence or if the accused had a fragile mental condition. In civil proceedings, inadmissible evidence includes, among other things, giving an opinion that the witness is not qualified to hold. For example, the vicar could not say that 'the crane driver was creating unnecessary noise by dropping the load from too high a height' unless he has professional knowledge of how cranes ought to be operated.

The rules on admissibility are extremely complicated, particularly in criminal cases, when a detailed knowledge of police procedures and the rules of criminal evidence is required to ensure that proper evidence is put before the court. Anyone charged with a criminal offence should obtain legal advice and representation immediately or risk allowing unsatisfactory evidence to be put to the court. This can make the difference between a guilty and not guilty verdict.

Since February 1997, new legal rules have been introduced with regard to the admissibility of 'hearsay' evidence in civil cases. Hearsay is indirect evidence or evidence of what someone else has said is true. It includes showing a witness statement to the court at trial without having the witness there to confirm orally what is written. Before the new rules were introduced, it was only in certain prescribed situations that hearsay evidence could be put before the court. Now, parties can chose to put forward hearsay if they think it will help their case. However, as with any evidence, hearsay should be judged in terms of how convincing it is and in most cases direct evidence is still preferable. So, the vicar telling the judge that he saw the crane driver at 6 a.m. is better evidence than the vicar's wife telling the judge that her husband told her about it. The wife, after all, cannot be certain that it did actually happen and cannot describe the noise.

PREPARING EVIDENCE IN CIVIL CASES

Although some of the rules of evidence are complex, in practice how the case is built is really a question of common sense. A lawyer starts with the legal definition of the law that his/her client alleges has been broken by the defendant. This could be the legal definition of the 'tort' (or 'wrong') of nuisance or negligence, or a breach of a duty defined in statute. The lawyer then begins to piece together evidence that will prove each part of the definition.

Why pre-action investigations are needed

Before starting court proceedings, a lawyer will undertake quite considerable evidential investigations. This is necessary for several reasons.

First of all, how a case is funded will depend on how a lawyer assesses the prospects of success (see Chapter 3). Crucial to this assessment is the quality of the evidence that is likely to be available. It is not possible to obtain legal aid funding unless there is at least a 50 per cent chance of the claim succeeding. This is so, even if the applicant for legal aid is on a sufficiently low income and has so few savings that the financial criteria for legal aid are met. At first it may be impossible to assess the chances of winning, and the legal aid granted could be limited to investigating the case further. However, unless the investigations uncover sufficient evidence to raise the prospects of success to over 50 per cent, the Legal Aid Board will not agree to fund the commencement of the legal action.

It may be possible for the case to be funded through a conditional fee agreement ('no-win no-fee') or by private means. Once again, however, an assessment of the prospects of success has an immediate effect. The lawyer will usually only agree to taking the case on under a conditional fee agreement if it is thought that there is a good chance of succeeding. Otherwise the lawyer risks not being paid at all. If the lawyer does agree to take on a risky conditional fee case, it is likely that the 'success fee' will be set at a high level (see Chapter 3, pages 57–61).

If the court case is funded through private means, the plaintiff risks being exposed to high legal costs. It is therefore well worth investing in the initial investigations.

A second reason for investigating the case in some detail before starting court proceedings is to enable the lawyer to advise on how much could be won in compensation or, for example, the likelihood of obtaining a particular court order. A lawyer cannot decide this without evidence.

A final reason for extensive pre-action investigations is a strategic one. The more prepared the case is the greater advantage there will be over the opponent when the case is begun. Proposed future changes in the civil case procedures will also favour the well-prepared litigant as court timetables look set to shorten the time allowed to parties to run a case, and those not ready with their evidence on the due date could find their case seriously weakened.

Steps taken in pre-action investigations

The plaintiff's lawyer will ask for any real and documentary evidence in the plaintiff's possession, custody or power and whether any witnesses are known who are prepared to help win the case. They will take a detailed statement to see how relevant, precise and convincing the plaintiff's evidence is.

The next step will be to find new evidence. Experts will be instructed to prepare reports to confirm whether, for example, the employer should have provided better health and safety equipment. Action may be taken to preserve evidence that could be destroyed or over which there is no control. This means writing to the relevant places and obtaining the documents or asking them to be retained. Medical records, for example, are frequently destroyed by hospitals after a certain period.

In some cases, access to the desired evidence may be refused. This frequently happens if the person or institution from whom the evidence is requested is a potential defendant. In cases involving an injury to mind or body, it is possible to apply to the court for an order against the holder of the documents that they be released. This is termed pre-action discovery, and is dealt with in more detail in Chapter 12, page 305.

Searching for evidence can be a question of following clues and being imaginative. For instance, a case may involve the development of mesothelioma, a rare asbestos-related disease that can lie dormant for forty years. The sufferers know that they worked with asbestos some thirty years previously. In order to win a case against the former employer (likely to be a firm/company *not* a person), the employer will have to be located, the conditions of work reconstructed and witnesses traced. Clues could be found in newspaper coverage of the time, adverts can be placed in the local press to bring old work colleagues forward, public records can be researched.

Anyone thinking of commencing a court action or who is at the investigative stage should ensure that any evidence thought to be relevant to his/her case is retained. If a dispute is ongoing, then records should be kept of when incidents occur, who spoke to whom, at what time and who was present. The record should make it clear how the potential litigant's life has been affected, including what he/she cannot now do as compared with what he/she could do before. Videos or photographs should be taken, particularly if a situation could change before an expert or a lawyer inspects the site or if it involves an activity. Damaged property should be photographed or kept. If money is spent, receipts need to be retained. If it is not clear whether an incident or a document is relevant, it should still be recorded or the document kept just in case. The lawyer can decide whether it will help.

Disclosing evidence to the defendants

Once a case has started, civil proceedings follow a traditional timetable of steps, which sets out when particular types of evidence need to be sent to the other side or to the court. This is described in Chapter 12.

Both parties to a court case are under an obligation to let the other side know what relevant documentary or real evidence they have. This includes documents that are unfavourable to the plaintiff's case as well as those that support it. Some documents, such as letters from lawyers advising on how good the case is, do not need to be disclosed.

This disclosure is known as 'discovery' and takes the form of a list of documents and other real evidence which is sent to the defendant. Witness statements and experts' reports are sent at a later stage. (For more on discovery, see Chapter 12, pages 303–6.)

If such a step has not already been taken, the lawyers should be provided by their clients with all the evidence that has been obtained or received since the lawyers were first consulted. First, the lawyer needs to be aware of all the evidence in order to decide how best to run the case. Further, it is strategically far better to serve a complete a list of evidence at the time set by the court timetable. The obligation to disclose evidence will continue until the end of the case and the court will not look kindly on piecemeal discovery in numerous lists unless there is a reasonable excuse. Moreover, the delays could be used as an excuse by the defendants to delay their own discovery.

Discovery in complex civil environmental litigation plays a particularly vital part in building up enough evidence to bring a case. This is mainly because of the difficulty in proving that a particular pollutant caused a specific illness or injury. Since 1992, those persons searching for additional data in environmental pollution cases can request information from institutions that carry out public administration, such as government departments, local authorities or regulatory bodies. Examples of these are the Environment Agency, the Health and Safety Executive and the Agricultural Research Council. Although information can be withheld for prescribed reasons (to be given in writing), these institutions can be a valuable source of independently researched data.

Another major problem to be encountered in the complex environmental pollution cases is the cataloguing and scrutinising of the evidence that is obtained. This is due partly to the sheer volume of material. For example, in the case against British Nuclear Fuels concerning childhood leukaemia around the Sellafield plant, the defendant disclosed approximately half a million pages of documents in lists every three or four months from the start of formal discovery to shortly before the commencement of the trial. These documents all had to be carefully considered not only by the lawyers but by experts. In such cases the court timetable has to be varied to accommodate the practicalities of the situation.

Witness statements

Witness statements are normally sent to the other side a few weeks after the lists of documents are exchanged. In a large and complicated case this might well happen several months later.

A good witness statement will be detailed and precise. It should contain facts and not opinions. A lot of time and thought goes into drafting the statements because they will form the basis of what is said in court. If a witness contradicts his/her statement when he/she is in the witness box this will seriously undermine the credibility of the evidence. Also, technically, witnesses cannot give any additional evidence in the witness box that could have been included in their statement, although they can give evidence on matters that have arisen after the date of the statement. As a guiding principle, those signing witness statements should ensure that they are prepared to stand by every detail. They should tell their lawyers everything they can remember and the lawyers will decide what needs to be included. Although it is much better to be as precise as possible, if a particular fact cannot be remembered this is what the lawyer should be told.

Finding witnesses to prove crucial facts can be very time-consuming. Sometimes agents are employed to find people whose names and addresses have been lost or who have moved. The best witness is someone who is happy to help but not so partial that his/her evidence becomes implausible. It is possible to obtain a court order forcing unwilling witnesses to attend trial to give evidence, although of course it is not possible to order them to sign a witness statement. However, a reluctant witness is a loose cannon because, without a statement from that person, in court the plaintiff's barrister will not know in advance what is likely to be said and it is possible that very damaging evidence might be given.

Those who have persuaded friends or acquaintances to act as witnesses and sign statements should ensure that they keep in touch with them as it may be several years before the case actually goes to trial. For obvious reasons it is also wise to keep on good terms with them!

In environmental cases, neighbours and others affected by a particular pollutant could themselves be parties to the litigation. In this case there should be no problem in obtaining secondary verification of the facts. Care has to be taken, however, to ensure that differences between cases are not overlooked.

Experts' reports

Experts are asked to give their opinions on technical issues that are within their particular area of expertise. Their reports are usually sent to the defendant after witness statements have been exchanged so that the experts have the chance to read the factual evidence before finalising their report. Technically, an expert's duty is to the court as an independent professional advising on matters that neither the individual lawyers nor the court have the knowledge and training to assess without help. However, in practice, each side will usually have its own experts who support one version of the story.

The experts are chosen for their specific qualifications and experience as well as for the favourable nature of their reports. For example, a general psychiatrist examining a plaintiff with behavioural abnormalities confirms that these are probably due to exposure to mercury during his employment. The defendant's psychiatrist, a pre-eminent specialist in mercury cases, gives evidence that, in her opinion, this is a case of schizophrenia, which can easily be confused with the symptoms of mercury poisoning. The general psychiatrist who has no experience of distinguishing between the symptoms of mercury poisoning and schizophrenia may have serious difficulty convincing the court that his diagnosis should be preferred.

There is little that the ordinary litigant can do to help the process of finding supportive and convincing technical evidence to back up his/her case. Obviously, the clearer and more detailed is the factual evidence provided through witnesses statements, the easier an expert will find it to relate his/her technical knowledge to the specific situation. However, in virtually all cases it is the lawyer who chooses the experts and asks them to prepare their reports on the key legal and technical questions. A lawyer's input can be critical: however good an expert is, if asked to report on the wrong areas then his/her evidence will not help. A lawyer will need to assess where a particular expert's expertise ends to ensure that the opinions expressed are restricted to that area alone. A lawyer will also have to judge whether the expert is 'too favourable' or perhaps appears to gloss over the difficulties of the case. Such evidence may be easily refuted by the opponent's expert and be discounted by the judge.

As the area of environmental litigation is relatively new, it can be a particular problem finding an expert both with the relevant qualifications and with experience of court work and knowledge of how to format a report for legal purposes. Lawyers have lists of experts with whom they have worked in the past, and are reluctant to use a new expert without them being recommended by someone whose judgment they trust.

Often, in very technical cases, the case can become a 'battle of the experts' because there is no concrete 'true or false' answer to the scientific question. There may be different schools of thought, for example, on whether an underground petrol tank has exposed a child to benzene poisoning and caused the development of cancer. In such cases, it becomes all the more important for the plaintiff to have top specialists who are at the forefront of scientific research and who are aware of the most recent scientific publications that either support or discredit the school of thought favourable to the case.

The top specialists are, of course, few in number and often extremely busy, travelling to various parts of the world. The Legal Aid Board will only be willing to cover reasonable costs, and although the most preferred expert may be the Professor of Genetics at the University of California, if a well-qualified geneticist from England is available there may be difficulties justifying to the Board the costs of flying the American professor to England and putting him up in a hotel. The choice of expert is also subject to availability. Some may need to be booked a year in advance in order to undertake a week's work. A major reason why trials of cases with several experts are delayed is the difficulty in finding a time when all experts from both sides are available for, say, a six-week trial.

Evidence at trial

Civil cases will be tried in the first instance by a judge or, if on appeal, by three judges. The job of the judge is to weigh up the evidence put forward by each side, together with their legal arguments, and to decide whether the plaintiff has proved that it is more likely than not that the defendant has broken a particular law and can be held responsible for the consequences.

The judge will consider the documentary and real evidence and decide what is their significance. How witnesses and experts fare under cross-examination is a crucial factor in any trial. Witness and experts are first asked by their own barrister to tell the court about the evidence they have given in their statements or reports. Witnesses are allowed to refresh their memories by referring to their statements. The barrister will try to have the witness put over the evidence in the most favourable light, avoiding placing unwanted attention on any weaknesses. Then the witness is cross-examined by the barrister from the other side, who will try to undermine the credibility of the evidence given. An effective cross-examination is one that places witnesses or experts under serious pressure. They can become unsure of their judgment and start to hesitate or contradict themselves. Numerous cases have been won or lost on someone's performance in the witness box. In cases involving the 'battle of the experts', when it is a question of preferring one school of thought over another, lawyers will hope to have experts who have demonstrated already that they can perform well under cross-examination.

Settling before trial

Most cases settle before they are ever heard by a judge. In the build-up to trial the lawyers will be assessing the strengths and weaknesses of their case, how a judge is likely to view it and whether witnesses or experts are likely to do well in the witness box. There may be last-minute exchanges of documents and statements that alter the balance of the case. The better and stronger the plaintiff's evidence the more likely a case is to settle on better terms. If the evidence is weak, or if the lawyers are worried that the key witnesses will buckle under pressure in the box, then the plaintiff may be advised to accept a low offer because the risk of losing is just too high. However good the case there will always be a risk of losing. Lawyers will have lengthy discussions trying to find the best way of presenting the case and will often disagree in their assessment of just how good a piece or evidence or an argument is. The process of going to trial or settling can feel like gambling rather than justice.

EVIDENCE IN CRIMINAL CASES AGAINST PROTESTERS

A lawyer starts preparing evidence in criminal cases in the same way as they do in civil. The prosecution and the defence take the definition of the particular offence and gather oral, documentary and real evidence to prove each part of the definition 'beyond reasonable doubt' or to construct a defence that raises reasonable doubt.

The way in which evidence is gathered and disclosed to the other side is different, however, and is tightly regulated by laws and codes of practice aimed at ensuring that the powers of the police are not exceeded and the rights of the defendant are protected. As a result, the preparation of a criminal defence involves detailed investigation of whether the proper procedures have been adhered to and, if not, whether evidence can be put to the court at all. These are highly technical issues and, in virtually all offences that an environmental protester is likely to be charged with, it is crucial that a defendant seeks legal advice.

How evidence is treated by the Legal Aid Board is also different in criminal proceedings. The quality of evidence or an applicant's chance of putting forward a good defence will not determine whether he/she obtains legal aid. Instead, a successful application depends on the seriousness of the offence and the complexity of constructing a defence. The legal and technical issues are considered as well as evidential difficulties in tracing and cross-examining witnesses. Most offences for which a person risks imprisonment are considered worthy of legal aid, although the applicant must still be on a sufficiently low income with few savings. Many low-level offences with which protesters may be charged (such as offences under airport bylaws or under the Public Law Act) are not imprisonable and therefore there may be difficulties in obtaining legal aid.

The events from arrest to final trial and sentencing can be bewildering and defendants may feel powerless and unable to help their own case. Although it is far more difficult for defendants to participate in their case than it is for civil litigants, there are ways to improve the chances of a not guilty verdict. The basic rule is to record as much as possible about what happened when, who was involved and what was said; the more contemporaneous the record the better. In this way the defendant's evidence in court will be more credible and could provide the lawyer with ammunition for challenging prosecution evidence on the grounds of procedural irregularities.

Arrest and detention by the police

Chapter 10 gives a brief outline of what happens from an arrest to a case being committed to trial and what rights a defendant has to information, legal advice and remaining silent. The right to information and legal advice should be exercised from the initial arrest, and, if refused, this refusal should be recorded straight away or as soon as possible. The lawyers should be told of this as soon as they arrive.

The police have wide powers to arrest and detain someone who is suspected of having committed a criminal offence. Clearly, little evidence is required to exercise this power and the police will need to substantiate their suspicions before progressing to formal charges. If possible, and as soon after the arrest as possible, the defendant should take the names and addresses of witnesses who may assist the defence. If there is considerable time between arrest and being taken to the police station, then an account of what occurred as a set of brief factual points should be written down, including the time, the date, the people involved, the actions of the police and, if possible, the police officers' numbers. If a lawyer or other independent person is monitoring events, then he/she should be informed of the facts.

The police will also, of course, start to gather evidence at the time of arrest and then continue at the police station. The methods used by the police to obtain the evidence will directly affect whether it can be considered reliable and therefore admissible.

For example, a police officer at a demonstration may have obtained the name of a witness who says she saw the defendant prevent a landowner from pulling up a hedgerow by standing with others in a ring around it. The arresting officer may not have seen this take place and will therefore need to establish by other means and beyond reasonable doubt that this is what happened. It is not enough that the witness gives a description of the defendant. The police may arrange an identity parade or one of three other methods of identification that are permissible under the police codes of practice. The number of people in the identity parade, whether the witness was properly segregated from those partici-pating in the parade and whether the defendant's rights were explained before the parade went ahead will all determine whether the identification was procedurally correct and therefore reliable.

It is not the aim of this chapter to explain all the rules governing police procedure and criminal evidence. Unless some-one has been arrested on frequent occasions, he/she is unlikely to have the presence of mind to remember detailed rules about how, for example, police interviews should be conducted and when a line of questioning may be considered threatening by the court. However, it is difficult to overstate how important it is that anyone arrested and taken to a police station should immediately exercise his/her right to legal advice. Organisers of a demonstration or protest that may lead to arrest should have the name of a lawyer who is known to be prepared to assist. Even if the offence with which a protester is charged makes it unlikely that legal aid will be available to pay for a lawyer, some sympathetic lawyers may be willing to help voluntarily. Once the lawyer has arrived it is his/her job to ensure that all rules are adhered to and to keep a written record if any breach of the rules is witnessed. A defendant will then have a much better chance of persuading the court that certain evidence was improperly obtained.

The Criminal Justice and Public Order Act 1994 has undermined a defendant's 'right to silence'. New rules now allow the court to draw adverse inferences from a failure to answer police questions during interview or a failure to give evidence at trial. In effect, this has started to shift the burden of proof onto defendants to prove their innocence. A lawyer will discuss what is the best way to respond to questioning with the client before and during any interview by the police.

From charges to trial

If, after detaining a suspect for the allowed time, the police believe that they have obtained enough evidence, then the suspect will be charged. It is the Crown Prosecution Service (CPS) who investigates and runs cases begun by the police.

The procedure by which a case comes to trial will depend on the type of offence with which the defendant is charged. Briefly, in summary offences (minor offences tried in the magistrates' court), the prosecution will inform the defendant of a trial date by letter. Barring applications for bail (required if the police refused bail after making formal charges) and adjournments to allow the defendant to prepare fully, then there will be no other court hearings. In offences triable either way (that is, in either the magistrates' court or the Crown Court), the first hearing concerns the 'mode of trial'. It takes place before the magistrates and decides whether the case will be transferred to the Crown Court. If transferred, then the either-way offence follows the same procedure as offences triable by indictment only (that is, only in the Crown Court). First, the prosecution must show through the committal procedure that the defendant has a case to answer. Then, a short hearing takes place at which the defendant must enter a guilty or not guilty plea and, usually, a date is fixed for the final trial. Sentencing after a guilty verdict takes place at the final trial or at an adjournment.

It is open to the defendant's lawyer to ask for an adjournment in order to prepare their evidence fully for any of the above hearings, provided that the request is reasonable. The court will want to been seen to allow sufficient time for the defendants to prepare their case, particularly when they risk imprisonment. However, strategically, the longer any delay the more time the prosecution is given to prepare its case.

Whatever the charges, the lawyer will start by taking a detailed statement from the client in private, regarding the circumstances of the alleged offence. The lawyer will want to iron out any inconsistencies and weaknesses quickly and as much detail as possible should be provided. Names of potential witnesses will also be needed. Delays in providing this information can be costly. Witnesses at the scene may not know the defendant and as time passes may be less willing to help. Events leading up to the arrest can be chaotic and fast moving. The exact sequence of events or the identity of those involved can be quickly forgotten. So, although it is usually open to either side to order a reluctant witness to attend court, if that witness cannot give a clear account of what happened then his/her evidence will hold less weight.

A defendant should give the lawyer the names of friends or colleagues who will act as a witness to his/her good character. Often, the court must decide which version of a sequence of events is the more believable, the prosecution's or the defendant's. A good character reference will help the defendant's credibility. It should be remembered that the defendant's evidence may be in direct contradiction to that of the police officer who was also at the scene. However, evidence of good character must be treated with care: in some circumstances the prosecution can use it as an opportunity to bring up a defendant's criminal record, which would normally be withheld.

The next step is to obtain evidence from the prosecution. In the past, in contrast to civil proceedings, the defendant was not under a duty to let the other side know on what evidence he/she would be relying, although any alibi had to be disclosed. However, since the Criminal Prosecutions and Investigations Act 1996 came into force on 1 April 1997, a whole series of disclosure regulations now apply. Defendants must now give an outline of what their defence is, in advance. In addition, defendants should ensure that their own lawyers have full information, including any information that may be prejudicial to them. The lawyers will then be better equipped to construct a clear and detailed defence and prepare for matters that the prosecution may raise.

If charged with an either-way offence, the defendant has the right to see the prosecution evidence in advance in order to prepare for the mode of trial hearing. In summary offences there is no right to advance information but the prosecution may be willing to release it anyway. The advance information should contain evidence that supports the defendant's case as well as the prosecution's. It will include written statements by the prosecution witnesses, documentation of the police procedures used to obtain evidence, and a list of a defendant's past convictions, together with the past convictions of those being tried at the same time. This information should be gone through carefully with the lawyer to determine what is accurate or inaccurate and what is remembered about each of the prosecution witnesses.

The defence lawyers will consider the advance information in great detail, looking for weaknesses and checking for procedural irregularities. They may want to talk to some of the prosecution witnesses, particularly if the prosecution has left stones unturned. The fundamental principle is that a witness does not 'belong' to either side.

Expert evidence can play a role in criminal proceedings if there are technical matters that the court cannot decide without a professional's opinion. This is rare when dealing with the type of offences with which public protesters are generally charged. Far more time is usually spent on establishing what actually happened at the scene of the alleged crime. A defence lawyer may visit the site and make plans and take photographs. It is a good idea for defendants to do the same as this may serve to jog their memory about what happened. For example, the results of a site visit (plans and photos and oral evidence) could help convince the court that the prosecution witness who allegedly identified the defendant in an identity parade was standing too far away to have a clear view in the midst of a throng at dusk.

Prior to trial, the defence lawyers will decide with the person charged what is the best way to run the case. In principle, the defence does not have to put any evidence forward. If the prosecution evidence is so weak that it is unlikely to lead to a conviction, then the defence lawyers may not want to run the risk of putting a witness into the box. The pressure placed on witnesses in criminal cases is the same as that in civil. However, the new rules undermining a defendant's right to silence will play a part in planning the case strategy. If the prosecution evidence raises issues that call for an explanation from the defendant that he/she then fails to provide, a court can infer guilt from that failure.

The rule against hearsay has not been abolished for criminal proceedings but there are exceptions to it. In the time leading up to trial each party can serve formal notices to enable them to rely on documents or statements that will not be authenticated by the makers of the evidence at trial. In all likelihood, however, the other side will only allow undisputed hearsay evidence to be put forward in this way because they will want the opportunity to cross-examine the witness.

Trial

If a defendant has pleaded not guilty then a full trial of the evidence will take place. In the magistrates' court, the magistrates are judge and jury. As jury, they decide whether the prosecution has proved beyond reasonable doubt that the person charged has committed the offence. As judge, they decide whether a piece of evidence should or should not be put to them. If considered inadmissible, the evidence is ignored for the purposes of the trial. Clearly this is an unsatisfactory and artificial procedure. In the Crown Court, the judge alone will hear arguments from lawyers on whether evidence can be put to the jury and the jury decides whether the prosecution has proved its case.

If asked to by the defence, the judge or magistrates may hear representations from the barristers on whether there is still a case for the defendant to answer. If the prosecution has failed to provide sufficient evidence on an essential element of the offence or if its evidence is so fundamentally unreliable that there can be no safe conviction, then the case is dismissed.

The prosecution is under a duty to ensure that all material evidence supporting the prosecution case is put before the court in a dispassionate and fair matter and it will be up to the defendant's lawyer to point out if this has not been done. The defence can challenge the admissibility of prosecution evidence at any stage. Frequently these challenges are made on the grounds that a witness of facts has given opinions, or that evidence is hearsay or 'self-supporting' (someone cannot, in principle, give evidence that they previously told the same story).

The barristers from both sides will present their witnesses' evidence and cross-examine the other side's evidence in the same way as in a civil trial. They cannot ask 'leading questions' or questions that suggest the answer expected, or assume the existence of a fact that has not been proved. A good barrister will instead pose a series of questions that elicit the wanted response from the witness. The judge or magistrates may also ask questions of witnesses about their evidence.

Ending the case before trial

The CPS can decide to drop charges at any stage before final trial. They may be persuaded to do so by a defence lawyer who has sufficient evidence him/herself to raise a reasonable defence or to seriously undermine prosecution evidence. It is therefore in the interests of the defendant to let the lawyer have as much information as possible straight away.

CONCLUSION

This chapter is just a basic guide on evidence. There are numerous
detailed rules that have not been referred to. Although how a case is
built often comes down to common sense, those experienced in running claims or campaigns can provide invaluable advice. The golden
rule for anyone considering civil legal action or a private prosecution
is to start recording incidents as soon as possible. At the initial stages
the lawyer will have to rely heavily on what the client says and the
evidence provided. If charged with a criminal offence, it is vital to
record as much as possible and obtain legal advice immediately.

GLOSSARY

affidavit A written statement that is sworn on oath and used as evidence.

causation The relationship between cause and effect.

civil action Court cases brought between individuals or groups in society, not involving criminal offences.

common law (also **case law**) Law created by judges through their decisions as contrasted to statute law.

conditional fee agreement A 'no-win, no-fee' method of funding some cases. If the case is lost the solicitor is not paid at all. If the case is won then the solicitor is paid a 'success fee' from his/her client's damages.

conviction The finding of a person guilty of a criminal offence at trial.

criminal proceedings A court case normally brought by the state against someone who commits a criminal offence.

cross-examination Where a witness is questioned by the opposite party when giving evidence in court.

damages Compensation money for loss or injury suffered awarded to a party who wins a court case.

defamation Verbal (slanderous) or written (libellous). Publication of a false and derogatory statement regarding another person.

defendant A person against whom a civil action is brought; also a person being charged with a criminal offence.

duty of care The duty to take reasonable care not to cause harm to others.

either-way offence An offence which can be tried either in the magistrates court or in the Crown Court.

evidence The means by which the facts of a case are proved or disproved. Includes oral testimonies, documents or objects.

examination-in-chief Where the party calling the witness questions the witness in court.

ex parte An application in a court case made by either (1) an interested person who is not a party to the case or (2) one party in the absence of the other.

expert report The evidence to be given by an expert in the form of a report.

indictable offence Triable only in the Crown Court.

injunction A court order by which a person is ordered to do something or refrain from doing something.

interim relief An order made before the final trial, intended to last for a limited period only – for example, a part-payment of damages.

judicial review An application to the High Court to review the lawfulness of a decision made by a public body.

leading questions Questions that direct a witness as to the response he/she should give.

legislation Law made by Parliament.

legal aid Scheme of government funding of legal fees for people whose income and capital fall below a certain threshold and where the case is considered deserving.

liable Responsible in law.

litigant in person A person who represents him/herself in court.

litigation Taking a case through the courts.

negligence Doing something that a reasonable person would not do or failing to do something that a reasonable person would do.

nuisance An unreasonable use of property that interferes with another person's reasonable enjoyment of property.

plaintiff The person bringing a civil court case.

pleadings Written statements exchanged between the parties in a civil case which set out the legal and factual issues to be decided a trial.

pro bono With no charge.

prosecute To bring criminal charges.

public nuisance An act that interferes with the enjoyment of a right to which all members of the community are entitled.

settlement The stage at which the parties in a civil case reach an agreed end to the case.

standard of proof The amount of evidence required for the case to be won. In criminal cases the allegation must be proved 'beyond reasonable doubt'. In civil cases the court makes its decision 'on the balance of probabilities'.

statute An Act of Parliament

statutory charge The Legal Aid Board's claim over all money or property won by the legally aided party in a court case. The charge is used by the Legal Aid Board to recover from the winnings any of the Board's costs that are not to be paid by the losing party.

statutory duty A duty imposed by statute law.

statutory instrument Secondary legislation created by a body delegated to pass such legislation by an Act of Parliament.

statutory nuisance Things that are 'a nuisance or prejudicial to health' as defined by the Environmental Protection Act 1990.

sub judice Once a case is set down for trial and until conclusion of the trial, rules apply to prevent media coverage.

summary offence Offence dealt with by magistrates only

tort A civil wrong for which the remedy is an action in common law for damages.

witness statement A written account of factual events signed by a witness.

List of Contacts

ENVIRONMENTAL GROUPS

Set out below, in nothing like an exhaustive list, are some of the key environmental organisations based in the UK. They are all well worth contacting and are generally happy to give whatever assistance they can to new groups.

ALARM UK

9/10 College Terrace, London E3 5AN. Telephone: 0181 983 3572.

ALARM UK was formed in 1991 and exists to assist local groups who are campaigning against road schemes. It advises groups on campaigning techniques and provides them with transport information.

In future, the group will continue to campaign for more sustainable transport policies, give assistance to local groups, build up strong regional groupings and develop work on roads and the economy.

ASH

Devon House, 12–15 Dartmouth Street, London SW1H 9BL. Telephone: 0171 3141360.

ASH is a charity that aims to eliminate the single largest preventable cause of death and disease in the UK by influencing policy and public opinion on tobacco use. ASH investigates and advises on policies for tobacco control, making submissions to the government and giving evidence to parliamentary committees.

In 1997, ASH stepped up its campaign to protect children from tobacco by lobbying politicians on the need for tobacco control legislation such as bans on tobacco advertising and controls on smoking in public places.

Communities Against Toxics (CATS)

PO Box 29, Ellesmere Port, South Wirral L66 3TX. Telephone: 0151 339 5473. Fax: 0151 201 6780. Email: cats@gn.apc.org.

CATS is a national grassroots network of communities affected by waste-disposal facilities, contaminated land and chemical plants. There are currently over 100 groups affiliated to the network.

Council for the Protection of Rural England

Warwick House, 25 Buckingham Palace Road, London SW1W OPP. Telephone: 0171 976 6433. Fax: 0171 976 6373. Email: CPRE@gn.apc.org. Web site: http://www.greenchannel.com/cpre.

The Council for the Protection of Rural England was established in 1926. It has more than 45,000 members and has branches in every county. It campaigns for a more sustainable use of land and other resources in both town and country, and has recently been involved in campaigns for sensitive urban regeneration, together with its more traditional areas of concern such as town and country planning, and rural conservation and protection. It is very influential at a governmental level.

Cyclists' Touring Campaign (CTC)

Cotterell House, 69 Meadrow, Godalming, Surrey GU7 3HS. Telephone: 01483 417217.

CTC is a national cyclists' organisation providing advice on cycling issues and offering campaign support.

Earth First!

PO Box 9656, London N4 4TY. Telephone: 0171 281 4621.

Earth First! is a network of independent groups who believe in doing just that – putting the earth first. It is a non-hierarchical organisation, whose aim is to use direct action and to empower individuals as a part of confronting the ecological problems facing our planet. Earth First! have been active in numerous campaigns, in particular opposing road developments, but also many other forms of direct action.

Friends of the Earth

26/28 Underwood Street, London N1 7JQ. Telephone: 0171 490 1555. Fax: 0171 490 0881. Email: info@foe.co.uk.

Friends of the Earth is one of the UK's leading pressure groups. Since its launch in 1971 it has attracted around 72,000 members and formed 240 local groups throughout England, Wales and Northern Ireland.

There is also an international network covering more than 54 countries. It has a very strong local group movement, and these groups lobby local authorities and companies and are actively involved in a number of grassroots campaigns. Friends of the Earth works with all political parties and also lobbies the EU, the United Nations and international trade bodies.

Friends of the Earth has been heavily involved in road and transport campaigning, complaints to the EU in respect of the UK government's failure to implement environmental directives, and campaigns concerning acid rain, ozone depletion and climatic change.

In the near future, major campaign areas will include endocrine disrupters (gender-bending chemicals), waste minimisation and regulation, legal actions and the development of economic instruments of environmental improvement such as eco-taxation and liability.

The Green Network

9 Clairmont Road, Lexden, Colchester, Essex CO2 OLE.
Telephone: 01206 46902. Fax: 01206 766005.
Email: 100727.3110@compuserve.com.

The Green Network is a networking organisation, helping concerned parties in the field of human health and pollution to establish links with each other. The group works with professional scientists, farmers, consumer groups and sufferers whose conditions may be related to toxic chemicals. The group networks internationally, in the US, Europe, Pakistan, Malaysia and Japan among others, and is concerned with a wide range of chemical usage, including agricultural, industrial and veterinary products, and the chemical treatments of homes and offices.

Greenpeace

Canonbury Villas, London N1 2PN. Telephone: 0171 865 8100. Fax: 0171 865 8200/8201. Email: Info@uk.greenpeace.org. Internet home page: http://www.greenpeace.org.

Greenpeace is one of the largest environmental organisations in the UK with some 300,000 members. Worldwide there are about 4.5 million supporters in 158 countries.

Internationally, Greenpeace campaigns through non-violent direct action and lobbying against global warming, the destruction of the ozone layer and rain forests, and the development of the nuclear industry. It also campaigns to protect rivers and seas from pollution.

Its real strength is its ability to act, at speed, at an international level. It is seen by governments as being a major international force. Greenpeace has a policy of complete independence from, and non-affiliation to, any political party, business interest or other outside organisation.

London Hazards Centre

Interchange Studios, Dalby Street, London NW5 3NQ. Telephone: 0171 267 3387. Fax: 0171 267 3397. Email: *either* lonhaz@gn.apc.org *or* lonhaz@mcr1.poptel.org.uk.

The London Hazards Centre, founded in 1984, provides free advice and information on all aspects of occupational and community health and safety to individuals, trade unions, resident associations and community groups. Chief areas of concern are asbestos, the building industry, chemicals (including pesticides), musculo-skeletal disorders, occupational stress, sick building syndrome and safety legislation.

A help line operates on Monday, Tuesday, Thursday and Friday from 10.00 a.m. to noon and 2.00 p.m. to 5.00 p.m. An answering service operates at other times.

The National Society for Clean Air and Environmental Protection

136 North Street, Brighton BN1 1RG. Telephone: 01273 326 313. Fax:iimplications of pesticides. It has a comprehensive interest in all ational Society for Clean Air and Environmental Protection (NSCA) is a non-governmental, non-political organisation and charity.

Founded in 1899, the Society's objectives are to secure environmental improvement by promoting clean air through the reduction of air pollution, noise and other contaminants while having due regard for other aspects of the environment. The Society examines questions of environmental policy from an air quality perspective and aims to place them in a broader social and economic context.

At the present time the NSCA is looking towards regulations and the effect of enforcement of existing noise legislation, seeking better cooperation between the Department of the Environment, Transport and the Regions, the Home Office and environmental health officers in resolving noise problems.

The Pesticides Trust

The Euro Link Centre, 49 Effra Road, London SW2 1BZ. Telephone: 0171 274 8895. Fax: 0171 274 9084.

Email: peststrust@gn.apc.org.

YYThe Pesticides Trust is a science-based charitable public interest group, concerned with the health and environmental implications of pesticides. It has a comprehensive interest in all aspects of pesticides, including agriculture, health and safety, conservation, food and water.

The Pesticides Trust participates in the worldwide Pesticide Action Network and has an extensive programme of work in the UK, the European Union and in developing countries. The Trust works with the European Commission in promoting alternatives to pesticides and provides evidence to United Nations Regulatory Agencies on the incidence of hazardous pesticides worldwide.

Ramblers' Association

1/5 Wandsworth Road, London SW8 2XX.
Telephone: 0171 582 6878. Fax: 0171 587 3799.

The Ramblers' Association was formed in 1935 and today has over 110,000 members. The Ramblers' Association is a registered charity that seeks to help everyone enjoy walking in the countryside, to foster care and understanding of the countryside, to protect rights of way, to secure public access on foot to the open country and to defend the natural beauty of the countryside.

Reclaim the Streets

Reclaim the Streets can be contacted at:
PO Box 9656, London N4 4JY. Telephone: 0171 281 4621.

Reclaim the Streets is a radical direct-action group with a non-hierarchical structure, focusing on urban transport policy and its effect on society.

Road Alert!

PO Box 5544, Newbury, Berkshire, RG14 5FB.

Road Alert! is a grassroots direct action campaign targeting new roads.

Roadpeace

PO Box 2579, London NW10 3PW.

Roadpeace supports family and friends of road crash victims, campaigns for better enforcement of traffic law and changes to legislation.

The Royal Society for the Protection of Birds (RSPB)

The Lodge, Sandy, Bedfordshire SG19 2DL.
Telephone 01767 680551. Fax: 01767 692365.

The RSPB is the largest wildlife conservation charity in Europe with over 890,000 members. The RSPB purchases land and creates new nature reserves to provide habitats for birds and other wildlife. In the UK it manages one of the largest conservation areas, covering more than 93,000 hectares, including a wide range of habitats such as forests, lowland heath, wet grasslands, estuaries and reed beds. The RSPB undertakes research into conservation issues and actively lobbies central and local government together with industry and landowners to provide a better future for wildlife and the environment.

Surfers Against Sewage

The Old Counthouse Warehouse, Wheal Kitty, St Agnes, Cornwall TR5 ORE. Telephone: 01872 553001. Fax: 01872 552615.

Surfers Against Sewage (SAS) campaigns nationwide for clean seas on behalf of the 20 million people who use the British coastline every year. The increased interest in leisure activities and the development of wet-suit technology means throughout the year more and more people are spending an increasing amount of time in and on the sea.

Formed in May 1990, SAS has grown rapidly to become one of the UK's leading environmental pressure groups. High profile, media-orientated campaigning tactics, backed by reasoned debate, have resulted in SAS being described as 'some of the government's most sophisticated environmental critics'.

SAS campaigns for full treatment of all sewage discharged to sea, a complete cessation of dumping of toxic waste at sea, and for the greatest environmental benefit per pound of water company customers' money spent.

Transport 2000

Walkden House, 10 Melton Street, London NW1 2EJ.
Telephone: 0171 388 8386. Fax: 0171 388 2481.

Transport 2000 campaigns for a coherent and sustainable national transport policy that meets transport needs with the least damage to the environment.

Transport 2000 works for policies that will cut road traffic, improve rail and bus services, encourage walking and cycling, make maximum use of rail and water for freight and give greater priority to protecting the environment, saving lives and conserving national resources.

The major campaigns from 1997 include Streets Ahead for Local Authority Networks and Streets for People, empowering local communities to discourage car use and encourage alternative means of transport.

Women's Environmental Network

87 Worship Street, London EC2A 2BE.
Telephone: 0171 247 3327.

Women's Environmental Network campaigns on air pollution, transport and food miles, focusing particularly on women's issues.

WWF-UK (Worldwide Fund for Nature)

Panda House, Weyside Park, Catteshall Lane, Godalming, Surrey GU7 1XR. Telephone: 01483 426 444. Fax: 01483 426 409.

WWF-UK is part of the world's largest non-governmental organisation concerned with the conservation of nature and natural resources on an international scale. WWF has over 3.7 million members internationally and over 50 affiliated, associate and programme offices worldwide. WWF-UK has over 1 million active supporters and has spent nearly £8.5 million directly on over 2,000 UK conservation projects since its establishment in 1961.

WWF's mission for the 1990s is to achieve the conservation of nature and ecological processes by the preservation of diversity, by ensuring sustainable use of renewable natural resources and reducing pollution and consumption of resources and energy to a minimum.

Current UK campaigns include forests, fisheries and climate change.

NEWS AGENCIES

Press Association

292 Vauxhall Bridge Road, London SW1V 1AE.
Telephone: 0171 353 7440. Fax: 0171 963 7192.
Email: copypa@press.net. Web site: http://www.pa.press.net.

National news agency of the UK and Ireland.

Reuters

85 Fleet Street, London EC4P 4AJ. Telephone: 0171 250 1122. Fax: 0171 542 7921. Web site: http://www.reuters.com.

International news agency.

Associated Press

12 Norwich Street, London EC4A 1BP.
Telephone: 0171 353 1515. Fax: 0171 353 8118.

News and photo agency.

OTHER ORGANISATIONS

Department of the Environment, Transport and the Regions

Great Minster House, 76 Marsham Road, London SW1P 4DR.
Telephone: 0171 271 5000.
Web site: http://www.open.gov.uk-dot-dothome.htm.

Environment Agency

Head office: Rio House, Waterside Drive, Aztec West, Almonsbury, Bristol BS12 4UD. Telephone: 01454 624 400.
General enquiries: 0645 333111. Emergency Hotline: 0800 807060.
Fax: 01454 624 409.
Web set: http://www.environment-agency.gov.uk.

European Commission

Brussels at Rue De La Loi 200, B-1049 Brussels. Telephone: 00 3522 2991111.

European Information Office (UK)

Jean Monnet House, 8 Storey's Gate, London, SW1P 3AT. Telephone: 0171 973 1992. Fax: 0171 973 1900.

General Secretariat of the European Parliament

L-2929 Luxembourg. Telephone: 00 352 43001.

Planning Inspectorate

Tollgate House, Houlton Street, Bristol BS2 9DJ.

The Shell Fish Association of Great Britain

Fishmongers' Hall, London Bridge, London, EC4R 9EL. Telephone: 0171 283 8305. Fax: 0171 929 1389.

The Stationery Office

Publications: 51 Nine Elms Lane, London SW8 5DR.
Book shop: 49 High Holborn, London WC1V 6HB.
Telephone (enquiries): 0171 873 0011. Fax: 0171 873 8200.

Welsh Office Planning Department

Cathays Park, Cardiff CF1 3NQ.

Bibliography

Ball, Dimon and Bell, Stuart (1997) *Environmental Law*, third edition, London: Blackstone Press.

Casarett and Doull (1992) *Toxicology – The Basic Science of Poisons*, fourth edition. London: Pergamon Press.

Department of the Environment (1993) *Integrated Pollution Control: A Practical Guide*, London: Department of the Environment.

Department of the Environment (1997) *The United Kingdom National Air Quality Strategy*, London: Department of the Environment.

Department of Transport (1995) *Public Inquiries into Road Proposals: What You Need to Know*, London: Department of Transport.

Harding, Thomas (1997) *The Video Activist Handbook*, London: Pluto Press.

Judicial Studies Board (1992) *Guidelines for the Assessment of General Damages in Personal Injury Cases*, London: Blackstone Press.

Kemp, David, and Kemp, M.S. (1982) *The Quantum of Damages*, revised edition, London: Sweet & Maxwell.

Pugh, Charles, and Day, Martyn (1995) *Pollution and Personal Injury Toxic Torts II*, second edition, London: Cameron & May.

Road Alert! (1997) *Road Raging: Top Tips for Wrecking Roadbuilding*, Newbury: Road Alert!

Ward, Sue (1992) *Getting the Message Across, PR, Publicity and Working with the Media*, London: Pluto Press.

Index

Index by Auriol Griffith-Jones